THE EVOLUTION OF MAN'S CAPACITY FOR CULTURE

six essays by

J. N. Spuhler · Ralph W. Gerard · S. L. Washburn
Charles F. Hockett · Harry F. Harlow · Marshall D. Sahlins

and a summary by

Leslie A. White

arranged by

J. N. Spuhler
Department of Anthropology
University of Michigan

DETROIT · WAYNE STATE UNIVERSITY PRESS · 1965

Library of Congress Catalog Card Number 59–10223

Waynebook number 17

Paperback reprint, 1965

Second Paperback reprint, 1968

To Alfred L. Kroeber

perceptive student and persuasive teacher of the origin of man and culture

PREFACE

This small book presents some new ideas, demanded by new facts discovered or reinterpreted since the publication in 1928 of Alfred Kroeber's "Sub-human Cultural Beginnings" in the *Quarterly Review of Biology*. These new facts give strength to a current examination of the evolution of man's capacity for culture, or the interconnections between the origin of man and the origin of culture. We still cannot cover the whole of this subject without slipping over unknown areas with indefinite statements. But the areas in which we can be definite are much larger than they were thirty years ago.

Three examples of these areas of new information or new ideas may be indicated here; others are considered fully in the text to follow: The fossil evidence for one of the three extinct genera which are the better candidates for the more recent of human ancestors is new and the evidence for the other two is richer by many-fold. Systematic field studies on man's living primate relatives are new and the volume of laboratory studies on primates has more than doubled. Only within the last decade have we realized fully the influence of behavior on our present bodily structure—that our heads, brains, and faces reached their present shape following, rather than preceding, the making of tools.

The six essays presented in this book were delivered in the Plenary Session of the Fifty-sixth Annual Meeting of the American Anthropological Association, Chicago, Illinois, 29 December 1957, in a symposium entitled "The Evolution of Man's Capacity for Culture" and were first published in the February 1959 issue of *Human Biology*. The "Summary Review" by Professor White was written especially for the book.

J. N. Spuhler

University of Michigan
Ann Arbor, Michigan

CONTENTS

SOMATIC PATHS TO CULTURE

BY J. N. SPUHLER[1]

University of Michigan

FOR the present discussion, I am going to assume that culture is a biological adaptation, with non-genetic modes of transmission, which greatly supplements somatic evolution. Viewed in this way, there is a gap between cultural behavior and non-cultural behavior. The two sides of the gap are defined in terms of symbol and lack of symbol. Also, viewed in this way, we see that the gap is bridged. The gap was crossed in the past by the human species, and it is still being crossed by babies as they learn to become human.

Now certainly the behavior we observe in all human societies is fundamentally different in some respects from the behavior we observe in societies of monkeys, apes, and insects. We all recognize the rich symbolic character of human behavior. But I want to stress the bridge —the crossing of the gap from non-symbolic to symbolic behavior—and not the gap itself. To me it appears absolutely necessary to consider

[1] Aside from minor changes, the present paper is the version presented in the American Anthropological Association symposium on the " Evolution of Man's Capacity for Culture," Chicago, 29 December 1957.

both in the history of the species and of the individual that certain conditions are prerequisites for the full acquisition of culture whether by individuals or by the species. And we find these elements in babies before they begin to use symbols and we find certain of them in the behavior of monkeys and apes. Of course the above statements refer to culture in the specific sense of that variety of culture realized by members of the genus *Homo* and not to culture in the generic sense, not to all conceivable varieties of culture.[2]

Since all agree that modern man has culture, or is cultural, the easiest way in which I could discuss the morphological paths to culture would be to summarize the evidence which supports some particular phylogeny leading to man. The argument would then be: 1) Man has culture. 2) This or that phylogenetic diagram tells us how man changed from some non-cultural stem primate in the Paleocene to what he looks like today. In one sense this would be a proper discussion of the somatic paths to culture. It is important, and perhaps lucky, that we can make such pictures of man's biological history with a considerable amount of credibility. But this approach, while valuable, in not sufficient to help us understand the historical biology of human behavior, and I am not going to give it much attention here.

In thinking about human phylogeny, I believe in using all, or nearly all, the hominoid fossils we know about, so long as they are not fragments. To argue that none of the known man-like fossils are in *the* human phylogenetic line seems to me obscurantist. To argue that the fossils we know about are "somewhat near" but not exactly on the main line seems unnecessarily cautious and hedging and may give the unknown greater weight than the known. Perhaps Weidenreich (1946) and Heberer (1950) went too far in using all known hominid specimens they considered authentic. But I prefer their use of all of them to Wood Jones' (1948) use of almost none.

There is not space here to give a review of new developments in human paleontology. In the last few years we have acquired a wealth of new specimens and new ideas and we have also been able to discard some old specimens and ideas with good cause. If additional fossils become available it may be necessary to make major revisions in what I am about to say. By taking an abstract level—the level of

[2] In a more extended treatment it would be useful to make "society" a level of integration between biology and culture. There are a number of recognizable and important primate "social paths to culture."

the taxonomic genus—I can avoid some undecided issues on the phylogenetic placement of individual specimens. For the moment I am going to assume a human evolutionary sequence of 4 or 5 genera (Clark, 1955):

1) Leaving out the periods before the Miocene, we start with *Proconsul,* the earliest ape whose skull is known (Clark and Leakey, 1951). I assume *Proconsul* had precursors who developed the general features of a man-like thorax and arms as we know them today, but that these terrestrial apes were not highly specialized as brachiators. There is no reason to suppose that any human ancestors since the Miocene have been arboreal to the extent characteristic of living gibbons, orangutans, or chimpanzees (Leakey, 1952; Washburn, 1951).

2) We don't know what happened in the Pliocene.

3) At least by Early Pleistocene there is *Australopithecus,* now known from dozens of good, or as Broom would say, "beautiful," specimens, and the earliest evidence of man-like animals with bipedal locomotion.

4) By Early Pleistocene times, and lasting into Middle Pleistocene in parts of Asia, we have the genus *Pithecanthropus.* From the neck down they were very like the genus *Homo* and like him they were tool makes, fire users, and hunters. Their brain volume was intermediate between *Australopithecus* and *Homo.*

5) At least by Middle Pleistocene we have the genus *Homo,* represented by such forms as Swanscombe, Fontéchevade, the Neanderthals, and Upper Paleolithic man. Everyone agrees that some, if not all, members of the genus *Homo* have culture.

Now, in the context of this sequence of 4 known genera, and with comparisons from living monkeys and apes, I want to discuss 7 biological topics which are preconditions for the beginning of culture. They are:

1) Accomodative vision,
2) Bipedal locomotion,
3) Manipulation,
4) Carnivorous-omnivorous diet,
5) Cortical control of sexual behavior,
6) Vocal communication,
7) Expansion of the association areas in the cerebal cortex.

Of course, these 7 conditions alone did not make a population of apes lacking culture into a population of men with culture. The evolution of man was not predetermined by a few conditions in a population of Miocene apes. Mutations are the fundamental genetic events in the historical process of the acquisition of the capacity for culture. Mutations are random events that do not point in an orthogenetic direction. But mutations are limited by the structure of the gene which mutates and this structure is determined by the evolutionary forces, especially selection, active in the history of the gene. In this way populations that survive accumulate genes which are favorable in the prevailing environment of the population.

To illustrate the complexity of human evolution since the Miocene as seen at the mutational level, let me do some speculative arithmetic—using figures that have fair justification and are conservative (Spuhler, 1948, 1956; Simpson, 1953). From the Miocene to now there must have been at least two million generations in the hominoid line. If the total breeding population in successful phyla was 10 thousand, we have 20 billion individuals as real or potential ancestors of modern man. If genes at the average locus mutate at a rate of 1 in 100,000, and if only 1 in 200,000 of these result in new and favorable steps (and that is a low estimate), we still could have about 20 thousand "visible," favorable mutational steps (in all loci) since the Miocene in the hominoid line.

Thus when we talk about 7 conditions we are perhaps oversimplifying the matter. But there is not time for further discussion, even if we knew what to say. And, I should add, the 7 conditions I list do not represent unit mutations, although mutation is the ultimate source of the genetic variation in each condition. Further, the order of listing is not strictly chronological. Evolutionary changes in the 7 conditions were interdependent and roughly synchronous.

1) *Accomodative vision.* Vision has been the primary sense in vertebrates as far back as we know them (Polyak, 1957). It makes possible their great mobility. The most complex vertebrates, birds and mammals, interact with their external environment predominantly *via* their eyes. Under the influence of the arboreal habitat, primate vision was perfected into a leading sense. Visual behavior is one key difference between the nocturnal, mostly solitary Prosimians, and the diurnal, more social Anthropoidea. The difference between these two is perhaps the largest gap in non-human Primate social behavior. With upright, or sitting-up posture, vision in the Anthropoidea gained strict control of

manipulation—it became super*vision,* a guide and control of fine manipulation.

The relationship between the evolution of keen vision and fine manipulation is two-directional. As Polyak (1957) says: ". . . vision itself [became] more refined and the intellectual absorption and mental utilization more complete and lasting, as the skilled movements became more complex and more efficient." We will find that this kind of both-way causation with two or more systems evolving simultaneously, where progress in each stimulates change in the other, is important to the understanding of many topics in this symposium.

Before taking up bipedal locomotion, let me mention one good thing that came out of the Piltdown affair. It was the insight given, for example in Hooton's excellent paper of 1925, on the asymmetrical character of human evolution. Hooton was right, mostly for the wrong reason (Piltdown), but he was early to stress that different regions of the human body change at different rates. Many workers today would follow Washburn's (1951) separation of the human body into three regions distinct in phylogeny, with arms and thorax the oldest, the bipedal complex of pelvis and legs later, and the head and face latest of all to reach their modern form.

2) *Bipedal locomotion.* Although functional differentiation of the front and hind limbs started with the first tetrapods where the front legs reach out and the hind legs push, *Australopithecus* is the first primate with upright bipedal locomotion (the tarsiers are bipedal hoppers). The australopithecine pelvis, sacrum, and femur resemble modern man in those features which make his upright posture possible. There are some features of full bipedalism not found in *Australopithecus*—these are fully developed in *Pithecanthropus* from Java and Peking. Australopithecine locomotion was certainly more similar to that of *Pithecanthropus* and *Homo* than to any of the quadramanus primates. We must conclude that, by the early Pleistocene, hominoids were bipedal with free hands which could be used to handle tools. We will see that this was a master adaptation that demanded other adaptations leading to man's capacity for culture.

3) *Manipulation.* A good start toward precise manipulation is seen in monkeys. When monkeys sit up their hands are temporarilly free and are used to bring objects close to the organs of touch, vision, taste, and smell. But something like a quantum jump is made when the hands

are continually free for such activity as they are in an upright, fully bipedal hominoid. Then the arms and hands—under the guidance of binocular vision with good accommodation—are principal organs for interaction with the immediate physical environment. Getting food, eating, grooming, fighting, making, using, and carrying tools, these manipulations, accompanied by a rich flow of sense data including those from the more developed proprioceptive arm-and-hand muscle sense, enlarge the flow of information to the brain which in turn fosters development of association areas for storage of past experience with the hands and guides and initiates new hand movements. The neural delay required when some extra-organic tool is interposed between stimulus and response probably had much to do with the first ability to use symbols and the start of language.[3] The co-adaptation of the hands, senses, and association areas in precise manipulation seems a first basis for the subsequent development of human intelligence.

4) *Carnivorous-omnivorous diet.* Man and the tarsier are unusual among primates in being carnivores. Many monkeys are omnivores and take small animals as prey. Man is unique among living primates in taking large animals for food and these in large numbers.

Fortunately we have some fossil evidence on the problem of diet. It is still an open question whether the Australopithecines were hunters or the hunted. But by Middle Pleistocene times the *Pithecanthropus* of Peking were hunters of large mammals as well as gathers of hackberries and other plant food.

The change to a partially carnivorous diet had extremely broad implications for the social organization of early hominoids. Carnivores get a large supply of calories at each kill. This concentrated food is more easily transported to a central, continually used shelter than is low-calorie plant food, especially before containers were available.

Whoever killed the baboons and bucks associated with the Australopithecines must have been tool carriers as well as tool users. Tool carrying implies a degree of conceptualization not required in the occasional use of tools (White, 1942; Bartholomew and Birdsell, 1953). Before starting on the hunt there must be a minding which associates the tool with an event which is to occur in the future. This type of

[3] Probably no one today knows the exact significance of this. See C. Judson Herrick, The Evolution of Human Nature, University of Texas Press, Austin, 1956 for suggestions on the evolution of human mentation.

mentation has not been observed in captive chimpanzees or monkeys, and certainly not in wild non-human primates. The archaeological record shows it was a consistent part of *Pithcanthropus* behavior by Middle Pleistocene times.

Compact animal protein high in calories is a good basis for food sharing. Of non-human mammals it is only the carnivores that share gathered food. It is unlikely that the long dependency of human children—so important in the acquisition of culture by individuals—could develop in a society without food sharing. And the amount of information which needs to be transduced in a communication system for plant eaters like the gibbons is small compared to that needed in group-hunting of large animals. Gibbons share, by vocal communication, knowledge about the location of food collected and eaten individually on the site; hominoids share in the location, collection, and consumption of food.

5) *Cortical control of sexual behavior.* There seems little danger that modern anthropologists will overlook the importance of sex in the evolution of culture. Some of us fail to emphasize that, with regard to the physiology of sexual behavior, man is neither a) completely like most other beasts, nor b) completely different from non-human animals. Here, as in many other biological characters, the apes and man are alike and man and the apes are unlike other mammals. In the majority of mammals sexual behavior is seasonal and the sexual periods correspond to times when the female has high probability of ovulation and conception. In such mammals including the lower primates, copulation is evoked by an increase of gonadal hormones in the body fluids. In such animals we can bring about, or prevent, copulation by gonadectomy and hormonal injections. But in man and the chimpanzee, and probably also in others apes, copulation is strongly under cortical control and is not prevented by gonadectomy (Ford and Beach, 1951).

An important adaptation for culture is the change from built-in nervous pathways to neural connections over association areas (where learning and symboling can be involved) in the physiological control of activities like sleep, play, and sex. Cortical rather than gonadal control of female sexual receptivity may not be essential to the hominoid family (observations on other animals suggest not), but cortical dominance in sexual activity may have contributed to the easy transition of the family from a social unit where sex and reproduction were more important than food economy to a unit where subsistence is the dominant familial function.

6) *Vocal communication.* Human speech is an overlaid physiological function. It uses a set of body parts of quite diverse primary action. Consider the muscles used in speaking. Most of our coordinated muscular movement involves corrections and adjustments from proprioceptors. But the laryngeal muscles lack proprioceptors, and feedback control of speech comes by way of the ear and the 8th cranial nerve. When we talk, the voice box, tongue, and lips must work together smoothly and precisely. The 10th nerve controls the adjustment of the vocal cords and the 5th nerve the movement of the lips. Both of these involve branchial muscle while the 12th nerve moves the tongue with somatomotor muscle. The neurological basis of speech is not clear, but it is clear that the only place where the motor organs and steering apparatus of speech are wired together is in the cerebral cortex. Perhaps hand-tool manipulation in group activities like hunting coordinated by vocalization may have helped to make the connections.

Although the larynx is homologous in all primates its position in the throat differs in man. The larynx of quadrapedal primates from the lemur to the chimpanzee is in close to slight contact with the soft palate. This is why chimpanzees cannot make long, resonant sounds. As a consequence of upright posture and flexion of the craniofacial base, the larynx in man is moved down the throat away from contact with the soft palate, and an oral chamber is formed which makes possible resonant human phonation (Kelemen, 1948; von Bonin, 1955).[4]

This is not to deny a rich variety of vocal production to the chimpanzee and other primates. The position of the larynx, however, is one reason why attempts to teach chimpanzees English have failed. Unfortunately no one has tried seriously to teach a chimpanzee to learn to speak using chimpanzee "phonemes."

7) *Expansion of the cerebral cortex.* Current statements in the anthropological literature regarding the size of man's brain often involve misintepretations in one or the other of two directions. On one extreme, some investigators stress the fact that, compared with *mammals* in general, especially large mammals, man's brain is unusually large, both absolutely and relatively. For example, a 150 pound man has a three pound brain, while a 150 pound sheep has a one-quarter pound brain, and a 1500 pound cow has a one pound brain (Brody, 1945). On the other extreme, the stress is put on the conclusion that man's brain is

[4] This was written before the publication of E. Lloyd DuBrul, Evolution of the Speech Apparatus, Charles C. Thomas Publisher, Springfield, 1958.

indeed large, but not unexpectedly so. For example, when the log of brain weight in *primates* is plotted against the log of body weight, the slope of the regression line is steeper than it is among mammals in general (proportional to the 0.79th power of body weight in primates, the 0.66th power in mammals), and on visual inspection the plot shows—as log transformations often do—remarkably little scatter (see von Bonin, 1952, Fig. 2), suggesting that brain weight in modern man is just about what would be predicted given the general regression of brain on body weight in primates and a knowledge of man's body weight alone (von Bonin, 1945, 1955). But if we take 1345 gm as a brain weight typical for modern man, say of 60 or 70 kg body weight (Bailey and von Bonin, 1951), we find man's brain is significantly larger than the value of 1095 gm of brain for 70 kg of body, predicted by von Bonin's (1955) regression formula: log brain weight = 0.79 log body weight — 1.00. A conclusion which avoids both extremes might stress at least two reasons for man's large brain weight: a) about 80% of man's brain weight may be explained because he is a primate of large body size, and b) about 20% of man's brain weight results from an evolutionary increase in the relative size of hominid brains—resulting in a total brain weight which is vast compared with mammals in general, and is significantly large compared with primates in general.[5] One reason we have overstressed the size of man's brain, even among primates, is that the chimpanzee and gorilla have relatively small brains, especially for primates. Similar arguments suggest that the frontal lobes in man, while well developed, are not of extraordinary and unexpected volume compared with other higher primates.

The distinctive feature of the human cerebral cortex is not so much in overall volume nor in relative size of the frontal lobes, but rather in the way that the projection areas are connected with association areas, especially in the temporal lobes, and in the way the whole thing works. I want only to point to these gross anatomical facts; Doctors Gerard and Washburn will take up their interpretation for cultural behavior.

Rates of human evolution. In closing let me call attention to two sets of observations about rates of human evolution, one from paleontology and one from neontology. The first has to do with the rate of hominoid evolution as measured in genera per million years. Consider

[5] In the oral presentation of this topic, I put more emphasis than now seems justified on the thesis that man's brain is only as big as one would predict for a large and typical primate.

some sequence like this one (based on chronological data from Zeuner, 1954):

PERIOD	MILLION YEARS AGO	GENUS
Miocene	12 (25)	*Proconsul*
Lower Pleistocene	1	*Australopithecus*
Middle Pleistocene	0.5	*Pithecanthropus*
Upper Pleistocene	0.25	*Homo*

Using only known forms, without guesses about unknown ones, this sequence of 4 genera is close to a minimum one for hominoid evolution. Almost everyone would agree that these are good genera, although the taxonomic distance between *Pithecanthropus* and *Homo* is not so great as between the others. If one insists that additional genera must be put in (and probably there ought to be at least one more for the Pliocene) it will only strengthen the conclusion I want to draw. The point is that there has been an unusually rapid rate of hominoid evolution as measured in genera during the past 12 million years, and especially in the past million. This is apparent when we compare hominoid rates with those for horses, chalicotheres, and ammonites (table 1). Something has speeded up hominoid evolution. I would guess that selection (perhaps within-species or inter-group selection) for a new type of environment —a cultural environment—has a lot to do with it.[6]

TABLE 1

Rates of evolution in terms of genera per million years
(Data, except hominoids, from Simpson, 1953)

LINE	NUMBER OF GENERA	MILLIONS OF YEARS	GENERA PER MILLION YEARS *
Ammonites	8	160	0.05
Horses	8	60	0.13
Chalicotheres	5	38	0.13
Hominoids	4	12	0.25
Hominids	3	1	2.00

* For extant lines: Number of genera minus one per million years.

[6] C. H. Waddington has suggested a non-Lamarckian mechanism whereby variations in ontogenetic pattern initially brought about by environmental influence may, if subjected to strong selection, undergo genetic assimilation. See his The Strategy of the Genes, George Allen and Unwin Ltd., London, 1957.

Ontogenetic rates. Insofar as phylogenetic information can be deduced from observations on living animals, there exists among pri-

TABLE 2

Average duration of prenatal and postnatal growth periods and of life span in different primates (Schultz, 1956)

PRIMATE SPECIES	GESTATION (WEEKS)	MENARCHE (YEARS)	ERUPTION OF FIRST AND LAST PERMANENT TEETH (YEARS)	COMPLETION OF GENERAL GROWTH (YEARS)	LIFE SPAN (YEARS)
Lemur	18	?	?	3	14
Macaque	24	2	1.8–6.4	7	24
Gibbon	30	8.5	? –8.5	9	30
Orang-utan	39	?	3.0–9.8	11	30
Chimpanzee	34	8.8	2.9–10.2	11	35
Gorilla	?	9	3.0–10.5	11	?
Man	38	13.7	6.2–20.5	20	75

mates a general evolutionary trend to increase the duration of the main periods of the life cycle. The evidence is presented in table 2. In the great apes the gestation period is lengthened to at least 34 weeks, full growth is attained by the end of the 11th year, and animals in their 3rd decade are senile. In man the duration of the prenatal period has changed little, if any, from that characteristic of the great apes, but the duration of the period of postnatal growth has almost doubled, and the total life span has more than doubled. Man is not unique with regard to the gestation period, but he is specialized in the marked elongation of postnatal growth and the long postponement of the onset of senility. These human specializations are extremes of trends found to lesser degrees in the evolutionary history of other primates (Schultz, 1956).

Man, then, is not much different from other primates, especially the apes, in the general sequence of events from conception to birth. After birth, the ontogenetic pattern in man differs markedly from that of all non-human primates but differs in a direction forecast by the general trend of primate evolution. I would guess that this elongation of the life periods after birth is a consequence of physiological adaptation to the acquisition of culture. Culture is a biological adaptation with a non-genetic mode of inheritance depending on symbolic contact rather than fusion of gametes. It has greatly supplemented somatic evolution. In all known human societies, individuals participate in social systems whose members represent more than a single biological family in which all are connected (as the social insects are) by gametes from one parental set. No human family is a self-sufficient system of social action. Symbols rather than gametes make this so. It may be assumed that the genes controlling the growth cycle in man have been changed through selection to man's *human, cultural* environment.

LITERATURE CITED

Bailey, P. and G. von Bonin 1951 The isocortex of man. Illinois Monogr. in Med. Sci., *6*: Nos. 1-2.

Bartholomew, G. A. and J. B. Birdsell 1953 Ecology and the protohominids. Amer. Anthrop., *55*: 481-498.

von Bonin, G. 1945 The cortex of Galago. Illinois Monogr. in Med. Sci., *3*.

——— 1952 Notes on cortical evolution. Amer. Med. Assn. Arch. Neurol. and Psychiat., *67*: 135-144.

——— 1955 Toward an anthropology of the brain. Annals New York Acad. Sci., *63*: 505-509.

Brody, S. 1945 Bioenergetics and Growth. Reinhold Publishing Corp., New York.

Clark, W. E. le Gros 1955 The Fossil Evidence for Human Evolution: An

Introduction to the Study of Paleoanthropology. University of Chicago Press.

——— AND L. S. B. LEAKEY 1951 The Miocene Hominoidea of East Africa. London: British Museum (Natural History), Fossil Mammals of Africa, *1*.

FORD, C. S. AND F. A. BEACH 1951 Patterns of Sexual Behavior. Harper and Brothers, New York.

HEBERER, G. 1950 Das Präsapiens-Problem. In: H. Grüneberg and W. Ulrich, eds. Moderne Biologie, Festschrift zum 60. Geburtstag von Hans Nachtsheim, Peters, Berlin, pp. 131-162.

HOOTON, E. A. 1925 The asymmetrical character of human evolution. Amer. J. Phys. Anthrop., *8*: 125-141.

KELEMEN, G. 1948 The anatomical basis of phonation in the chimpanzee. J. Morphol., *82*: 229-256.

LEAKEY, L. S. B. 1952 The Environment of the Kenya Lower Miocene Apes. IIe Congrès Panafrican de Préhistoire, Livret-Guide, Alger, p. 77.

POLYAK, S. 1957 The Vertebrate Visual System. University of Chicago Press.

SCHULTZ, A. H. 1956 Postembryonic age changes. Primatologia: Handbook of Primatology, *1*: 837-964.

SIMPSON, G. G. 1953 The Major Features of Evolution. Columbia University Press, New York.

SPUHLER, J. N. 1948 On the number of genes in man. Science, *109*: 279-280.

——— 1956 Estimation of mutation rates in man. Clinical Orthopaedics, *8*: 34-43.

WASHBURN, S. L. 1951 The new physical anthropology. Trans. New York Acad. Sci., Ser. II, *13*: 298-304.

WEIDENREICH, F. 1946 Apes, Giants and Man. University of Chicago Press.

WHITE, L. A. 1942 On the use of tools by primates. J. Comparative Psychol., *34*: 369-374.

WOOD JONES, F. 1948 Hallmarks of Mankind. Bailliere, Tindall and Cox, London.

ZEUNER, F. E. 1954 Chronological tables. In: C. Singer, et al., eds., A History of Technology, Oxford University Press, Vol. 1, pp. xlviii-lv.

BRAINS AND BEHAVIOR

BY RALPH W. GERARD
Mental Health Research Institute
University of Michigan

SINCE I am not an anthropologist but a biologist, a physiologist primarily, I take it that I am to discuss not culture but the physical, the biological, substrate of behavior. This is a pleasant assignment because living things really matter only in terms of their behavior and how they modify the environment. The most primitive organisms and the most advanced are pretty much alike in their dynamic equilibrium, the intake, digestion, metabolism, and excretion of food. The details differ but the same substances pass essentially through the same steps. The processes of growth and reproduction also are vastly different in detail but still basically alike in the chromosome and gene machinery.

It is primarily in ability to interact with the environment that the big changes have come in evolution and, of course, man is the outstanding interactor. Man has succeeded enormously in that there is probably more living protoplasm of the human variety than of any other protoplasm; certainly man has modified the environment more than all other organisms combined and is the greatest catalyst that nature has produced so far. This is because of his behavioral complexities, his repertoire of interactions, particularly his ability to interact with his fellows and collectively to do things that an individual man could not begin to do. I should like first to consider with you the evolution of behavior, and primarily in terms of the mechanisms that make it possible.

Some kind of behavior is present in every organism; it must in some way be related to its environment. This may in effect be nothing more than chemical interchange, although every cell has some ability to rehabilitate itself, some responsiveness and, to some degree, can conduct and transmit. But this is at the cellular level, and perhaps the major difference between plants and animals is that in plants these primitive behavioral capacities have remained overwhelmingly at the cellular level. Animals introduced some new tissues, new structures, and began to specialize in sensation, in conduction, in response. This is the great step that animals made. Even this didn't really become exciting until animals stopped being sessile and were able to move around in their environment. Not just to move a little in a radially symmetrical manner, in which any direction is as good as any other; this was not too different from no movement.

The big change came when animals developed a direction, a head end, a longitudinal gradient, and bilateral symmetry. One end was then in front, faced the future, met the environment, got the bumps. The nervous system showed a progressive movement towards that end, with a centralization of the cells and neurons scattered throughout the organism. Then came the development of distance receptors, new kinds of sense organs that gave greater information about the environment the organism would encounter ahead as it continued crawling, rather than what was happening to its tender skin after it had already arrived. Along with movement came also a premium on speed and control, with comparable evolution of the conduction mechanisms and the nervous system itself.

Now, this great invention and development of the units culminated, at least along one line, with the origin of vertebrates. Earlier there was great improvement in receptors, conductors, and detectors. The human eye is sensitive to 10^{-8} candle power; the clam eye only to 10^{5}. That is a ratio of 10^{13} times in sensitivity, 10 million million times greater sensitivity to light in our eye than in the clam's. The human muscle gives its twitch in a few hundredths of a second; the sea cucumber muscle takes three hours—a factor of a million. The medullated vertebrate nerve can conduct at over a hundred meters a second; the primitive jellyfish nerve conducts a centimeter in a second or longer. Again the factor is a hundred thousand or that order. Another very interesting point is that, with directional behavior and higher organization, came irreciprocal conduction in the nervous system. In primitive nervous systems, messages can go in either direction across the junctions between separate neurones, for the synapses are symmetrical structurally and functionally. Later, synapses become asymmetrical and permit only unidirectional passage of messages and information through the nervous system. I would again like to emphasize, however, that all this was essentially completed with the arrival of the vertebrates, not of man, or of the primates, or even of the mammals. The hawk's eye, in some respects at least, is better than man's; the dog's smell is better than man's, the snake has a temperature receptor that only our latest snooperscopes approach for infra-red and temperature measurements, of .005° C. and so on down the line.

What happened after the units were developed? Further advance had to be in improved connections and interactions and in sheer number of units. Let me exemplify this with the development of radios. Can you remember back to the early triode? When the heterodyne and super-heterodyne circuits came in, the number of tubes in the radio multiplied —a good set was one with a dozen or two. With improved tubes, the tetrode, pentode, and now the transistor which functions on a different basis, the unit itself became much more efficient and good radio sets now need only half a dozen tubes. Moreover, these could be miniaturized, the circuits printed, and the whole instrument represents our present technological level of advance. In the case of the nervous system and animal behavior, the unit in frog or fish is essentially as good as that in man; the frog's nerve conducts as fast as man's at the same temperature, and so on down the line. Improvement from fish to philosopher, then, must depend either on better circuits, better interconnections between neurones, or simply on more neurons.

The circuits are certainly important and, although we are far from understanding the full detail of how they operate, we have a good general picture. In contrast, little parallel attention has been given to sheer number. In playing Twenty Questions, if the number of questions permitted is limited to, say, two instead of 20, or increased to 200, the elaborateness or complexity, the intellectual richness of the material which can be handled, shrinks or expands dramatically. The modern computing machines have essentially the same units, whether simple little ones or the great "—IACs"; the differences are in more and more units, which increase performance not merely in a summative fashion but in some kind of multiplicative one. It is rather like Eddington's lovely statment, to the effect that we once thought if we knew "one" we knew "two," for one and one are two; but we have since found we must learn a deal more about "and." A large organization composed of the same kinds of units, say human beings, develops differentiation of labor, organizational patterns, and is able to do a great many things that a smaller one cannot, simply as a matter of numbers. Well, as long as behavior flowers while the units remain essentially alike (despite marked differences between neurons in different parts of our own nervous system and, to some extent, between those in different species, there is no progressive improvement in them); there must be some relation between total number of neurons, size of nervous system, and behavioral capacity.

I am interested in the statment of Spuhler (chapter 1, above) that extrapolation in the primate series indicates that man's brain is bigger largely to the degree expected because he is bigger. Maybe this is all there is to it, and little spurt of special cerebral evolution has occurred; but then the simple size increase was fortunate, for certainly man's big brain is what makes possible man's human behavior. This can be pointed up in some detail. I remind you that the size of the motor area which controls various muscles does not relate to the size of the muscle but to the skill in using it. I strongly suspect that you could not teach a chimpanzee to speak chimpanese, let alone English, because he doesn't have large enough motor areas for his tongue and his larynx. The motor area for the human tongue is much larger than that for his whole leg. The human area for a tail is non-existent; but in the spider monkey, with a wonderfully prehensile and manipulatable tail, the tail area is huge.

Let me take the remaining time for a little more concrete guessing—for it is really that—as to mechanisms in the nervous system. I shall consider first some structural and then some functional ones. With evolution, with cumulative racial experience starting in the individual and getting fixed in the race, by the genetic mechanisms mentioned, certain basic patterns are laid down. Thus, the reflex arc is standard equipment. Later came mechanisms higher up the neuraxis: first those related to awareness in the mid-brain; then those serving emotion, pain in the thalamus or between brain, and pleasure still higher in the old forebrain; and finally reason, with its attributes on which we pride ourselves, in the new forebrain or cerebrum. Further, two aspects of consciousness probably depend on two kinds of mechanisms. The content of awareness is probably related to the specific paths from the thalamus to the cortex, which carry localized patterns of activity that are distinctive for the individual past experience of that particular organism, and the current pattern of sensory stimulation. The general level of consciousness, rather than its content—whether one is drowsy or alert or excited—however, depends on a non-specific activating system from thalamus to cortex and partly, I believe, on the adrenalin level in the blood, which increases as part of the general emergency response of an organism to a threatening situation.

Next, I must point out that a major decision was made by animals quite early in evolution when they adopted, so to speak, an alphabetic rather than an ideographic language. The nervous system is made up of relatively uniform units, which can be combined in various ways, rather than of highly particularized units, each one of which stands for something specific. The proof of this is the existence of synapses. The synapse can only be present because it is important *not* to have a message go through on an express track from one receptor to one effector. Otherwise, why break the nerve path and cause slowing, the chance of confusion, and all the rest? The synapse permits changeability, allows the units to connect now this way, now that way. This leads to the last point on structural aspects. I prefer to think not of afferent and efferent flows involving the nervous system, but of information coming in and instructions coming out. Every synapse is thus, in effect, a decision point. A message comes to it from the pre-synaptic fiber; does it go out over the post-synaptic one or doesn't it go out? That is the decision the nervous system makes—at a near infinity of places and times. As to how this decision is made, I will have a word in a moment.

The great organizational problem of the nervous system I suspect, is what we might call, in military terms, its "table of organization." To what extent, for example, are these decision-points scattered peripherally, so that there is effective local reflex control of parts of the system, and to what extent are they centralized, so that there is a general or a president whose word is relayed down through subordinate officers and out to the executive agents, the privates or workers. The degree of true centralization, I think, is going to prove the most interesting morphological and morphophysiological question to ask about the nervous system. It hasn't been asked; I suggest that it is very much worth asking.

Let us, finally, take a quick look at some of the physiological and functional aspects of neural mechanisms. Much attention has been paid to morphology, especially in relation to neural evolution, because records are morphological and because, in the development of any science, the morphological phase precedes the functional one. There have been few efforts to look at the functioning of the nervous system not in terms of where does it happen, which part does this or that; but in terms of what differences in physiological properties—either of the unit neurons or of the patterns of their synaptic connections—could be responsible for this or that performance. This is, of course, much too large, and perhaps esoteric, to present in the few minutes left, but some important items follow. First there is growing evidence (but no solid proof) that the nervous system works in a series of time frames, perhaps of a tenth of a second. Next, there is evidence that impulses reverberate—run many circles around a neuron loop—and that this may be involved in the process of fixing experience. Unless there is time for a message to reverberate many thousand times, a memory is transient and experience does not leave a permanent trace. Third, attention is related to threshold changes in primary receptors. A mother can hear her baby's cry through general noise because of a positive feedback to the ear itself. When an animal is conditioned to listen for a particular sound, the potential evoked by it in the ear is increased; if conditioned to ignore it, the reverse occurs. Fourth, are clues as to the processes of reason and imagination and of the introduction of the novel.

Let me now put a few of these things together. I warn you that this is pure guesswork but I believe it is at least an educated guess. Suppose differences between nervous systems were examined in terms of one or another of the properties that make for transmission across synapses, or for reverberation, or for synchrony, or for laying down memory traces. If memory traces are fixed more easily than the norm in some individual he might have a quiz kid kind of mentality. Everything that comes in would fix pretty easily and a tremendous amount of memory residue accumulate; but, by the same token, if things fix too easily there should be less flexibility. I think it is the case that people with phenomenal memories are rarely very imaginative or creative. With poor fixation, on the other side, would come presbyphrenia, failure of recent memory in the old person, and the unpleasant ability to go through complicated interactions and, at the end, forget all about them.

Look next at threshold fluctuations. It is established that the threshold for a given synapse is not constant, aside from actual manipulation, but shows erratic, unpredictable spontaneous changes. These depend on random events—which means, on such submeasurable and unanalyzed factors as electron movement, Brownian movement, and the like. Now, a certain amount of this is very fine, because if there were none every input would find the system (once the basic paths were established) exactly the same so its output would be automatic and routine and there could be no innovation. This is the situation for spinal reflexes and emotional reflexes, they are pretty stereotyped. But with a play of thresholds, the same stimulus may at one time send impulses along this path and at another time along that path, and so favor innovation and imagination. If fluctuation goes too far, coherence would be lessened and flight of ideas occur. Finally consider the phenomenon of synchronization. Neurons are known to have electrical beats, and large numbers of them beat in phase, as armies marching in step. When impulses come in, from the eyes particularly, the neurones begin to scintillate, to get out of step with each other, and the brain wave rhythm breaks up. Suppose there are differences in the firmness of synchronization. Too firm a synchronization should give what I might call "tubular attention"—a person with blinders on all sides who follows one path and cannot be diverted to see anything else. With too weak synchronization, neurons would get out of step too easily, and flightiness and inability to maintain attention would result. With excessive inhibitory action there should be decreased attention or actual depression; with too little inhibition there could be anxiety and even mania or epilepsy. These are some physiological variables that I am suggesting as important for behavior—another problem for the future.

What I have undertaken is just to point up ways in which modern neurophysiology (or neurobiology) can begin to understand the nervous system as the substratum of man's capacities, as an organ which can play the tunes of behavior. For the future, with the recently added tools for increasingly precise manipulation of the brain and measurement of behavior, progress should be rapid. Fine electrodes can now be left in place indefinitely, and the brain activity sampled or excited. Animals can administer shocks to their own brains, or inject them with drugs. More specific, "psychoactive," drugs are now available as biochemical instruments to manipulate enzymes and so alter these physiological properties of neurons—to say nothing of their therapeutic value for the mentally ill. Experimental evidence on the kinds of relations between neural mechanisms and behavioral patterns may thus soon be at hand and lead to sharper guesses than those I have put before you.

SPECULATIONS ON THE INTERRELATIONS OF THE HISTORY OF TOOLS AND BIOLOGICAL EVOLUTION

BY S. L. WASHBURN
University of Chicago *

THE purpose of this symposium is to consider the interrelations of the human animal and his culture. The point I would like to make in this brief paper is that much of what we think of as human evolved long after the use of tools. It is probably more correct to think of much of our structure as the result of culture than it is to think of men anatomically like ourselves slowly discovering culture. The way of life of early man (pre *Homo sapiens*) must have been very different from that of our immediate ancestors. It is not certain what limits their biology may have placed on their way of life, and therefore, Hallowell (1956) has suggested that it is better to speak of the early men as possessing "proto-culture," and keeping the word "culture" for the way of life of *Homo sapiens*. As Hallowell points out, when we are dealing with anatomically primitive men, it is uncertain whether their intelligence, capacity for language, art and social organization were like ours. Probably, the biological potential for culture of Wadjak man was the same as our own, but the biology of Pekin man placed quite different limits on his proto-culture.

If we want to relate the evolution of culture to biological evolution, a first requirement would be fairly complete records, which do not exist. The biological record is extraordinarily scanty, as Dr. Spuhler (chap. 1, this vol.) has already indicated. The archeological record is richer but it is difficult to interpret and there are great disagreements. One does not find culture or proto-culture, but only clues as to what the way of life may have been. I am going to indicate a possible interpretation of the sequence of tools, and then speculate on the relations of this sequence to the fossil men (fig. 1).

It is possible that pebble tools are the only tools surely present before the time of the second glacial advance. These simple forms may have been invented many times, and they are too crude to give indications of cultural contact. Subsequent to the pebble tools appear traditions of

* Present address: University of California, Berkeley, California.

manufacture, tools which were made according to such complicated plans that one feels that there must have been communications among the makers. The great biface and chopper traditions extend over huge

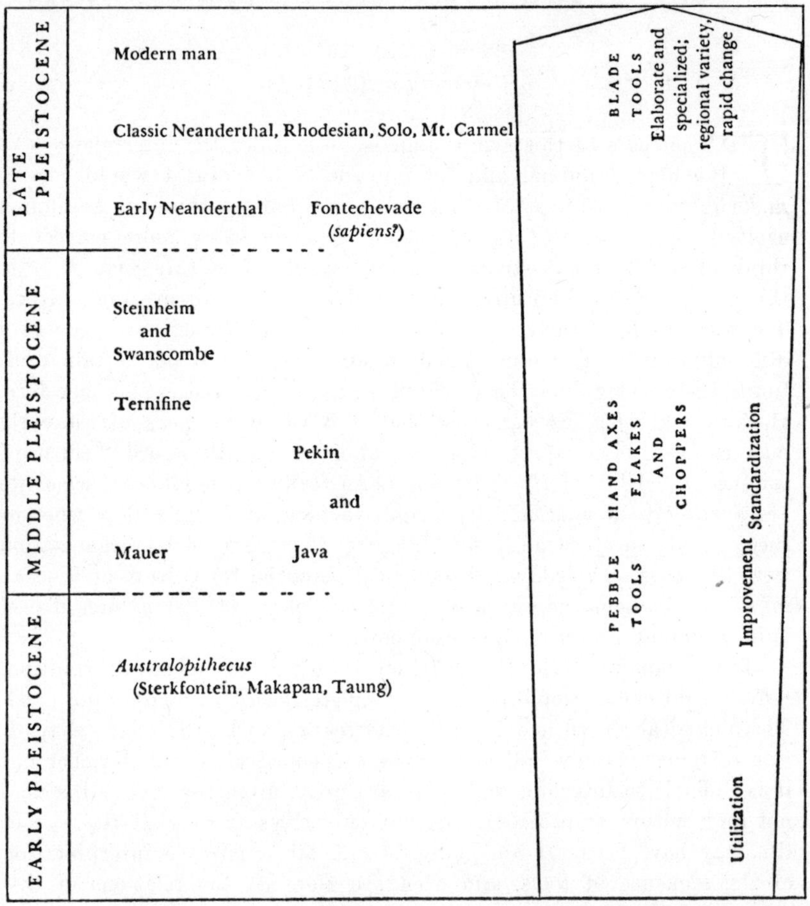

• Prepared with the aid of R. B. Braidwood, F. Clark Howell, and Kenneth P. Oakley.

Fig. 1. A Possible Correlation of Types of Tools and Fossil Men.

areas of the Old World, from second glacial times on into the last interglacial period. The persistence of these traditions is something which seems very strange to us today, for in subsequent layers the rate of change is vastly greater. Probably, the slow evolution of these proto-cultures (which we know only from the fragmentary record), is due in part to the bilological limitations of the early men. The rapid rate of change which appears during the time of last glacial advance is due both to the cumulative effects of culture and to the appearance of *Homo sapiens,* the biological form which made culture as we know it possible. As soon as *Homo sapiens* is surely on the scene, there is great regional differentiation. No longer does an Acheulian tool extend from Africa to India to Europe, and last for 100,000 years. Tool types change rapidly. Man crosses large bodies of water, conquers the Arctic, and art appears in the archeological record. This is the record of culture, the evidence of the presence of the restless creator we know as *Homo sapiens.*

According to this brief outline, there are three biological and cultural stages in the Pleistocene: an early-Pleistocene—Australopithecine—pebble-tool stage; a middle-Pleistocene—early man—biface, chopper stage; and an upper-Pleistocene—*Homo sapiens*—Upper Paleolithic phase. For the rest of this talk I want to speculate about this view of human evolution, to consider its implicaton, and to think about what kinds of evidence might help to prove, or disprove, it. As Spuhler (chapter 1, above) has indicated, man became distinct from the apes when he became a biped. The structure which made this change in locomotion possible is primarily in the pelvis and associated muscles, and it is because the innominate bones of four different specimens of *Australopithecus* all show a condition far closer to man than ape that these forms are regarded as bipeds. Since these small-brained bipeds of the early Pleistocene may have been the first tool users, I should like to be able to say something about their hands. Unfortunately none are preserved and as we all know, a whole hand is unlikely to be. However, it is not necessary to find a complete hand to make some deductions about its use. The finger bones are very different in the quadrapedal monkeys and in the arm-swinging apes. In the apes the phalanges are curved and the edges are highly developed. With a single bone it might be possible to give a highly-informed guess as to whether the australopithecines still used their hands for climbing frequently. It is not unreasonable to hope that such small fragments may be found. Further, man has much more motion in the wrist between the radius and the first row of carpal bones than monkeys do. The degree of motion is accurately reflected in the joint surfaces, and can be estimated from quite small fragments of bone. A Miocene ape in which the hand is well preserved, probably had motions of the hand and wrist much more like a monkey than like man. I suspect that a late Pliocene brachiating ape was the ancestor of the Pleistocene australopithecines, and that the structure of the hand which made tool use possible was evolved at that time.

It has been argued that ape hands are too specialized and different from those of man for a human hand to have evolved from the sort of hand seen in the living apes. Obviously, if our hand evolved from an ape hand, it was from a form long dead, and the hand may have been somewhat different from those of the living apes, but the essence of this argument is the assumption that the hand which first started to use tools was much like that of living men. However, if an early australopithecine with an ape-like hand had started to use tools, selection would have changed its descendants. Once these animals became bipedal ground dwellers, using tools a little, then selection would favor shorter fingers and larger thumbs. The use of tools would change the direction of evolution and the form of the hands. The essence of the interpretation of evolution by selection is that selection causes change, and I believe that the form of the human hand *is the result* of the new selective pressures which came in with the use of tools. According to this theory, an ape's hand was pre-adapted to the use of tools. The hand was freed by the assumption of bipedal locomotion. Then new selection pressures coming with the use of tools changed the ape hand into the human hand. Our hand is the result of at least half a million years of tool use. The uniqueness of the human hand, those features which distinguish it from the hands of apes, is the result of culture. According to this theory, it is futile to look for an ape-like ancestor with a large, fully opposable thumb, because the human thumb as it exists today evolved after bipedal locomotion and with the use of tools.

Turning from the hand to a more general view of human evolution, many people used to believe that men essentially like ourselves lived at the beginning of the Pleistocene. A common form of phylogenetic tree shows *Homo sapiens* as already in existence before the earliest tools. In such a tree all the anatomically primitive fossil men are put on numerous side branches. This led to attempts to relate the *sapiens* fossils to the biface tools and the anatomically primitive men to the flake tools. This type of correlation no longer is tenable because the supposedly early *sapiens* fossils have been discredited one by one. The separation of the flake and biface traditions was a great oversimplification. Undoubted associations of primitive forms of man and the Acheulian tools have been made. As Spuhler has indicated (chap. 1), it is now extremely unlikely that all the primitive forms belong on side branches. The general view of evolution outlined by Weidenreich (1947) is probably correct. According to Weidenreich, the men of the early Pleistocene were primitive in form, and *sapiens* evolved late.[1]

A possible evolutionary sequence is, the australopithecines of the early Pleistocene, Peking man and other primitive forms of the middle Pleistocene, and *sapiens* of the end of the Pleistocene. There is a clear difference between australopithecine and ape in the dentition, but most of the evolution of the brain comes after the australopithecine stage. It is my belief that the decrease in the size of the anterior teeth and the tripling of the size of the brain came after man was a tool user, and were the result of the new selection pressures coming in with the use of tools.

At Sterkfontein, Robinson and Brain have now found pebble tools. The implements and conditions of discovery have been examined by Desmond Clark, and there is no doubt that these are tools coming from the australopithecine levels and made of stone which does not naturally occur in the deposit. The evidence has been summarized by Oakley (1957) and he believes that the tools were made by *australopithecus*. I think that this is the most probable interpretation, but some think that no creature with so small a brain could have made tools and that more advanced forms of man must have made them, perhaps using them to kill the australopithecines. This sounds like the history of Peking man all over again. However, one cannot be sure, and, as pointed out to me by Dr. Robert Braidwood, this sort of controversy can only be settled by further careful excavation. An association in one site cannot settle the matter, and the situation is further complicated because several kinds of *Australopithecus* and *Homo* may have made tools. It is perfectly possible that different species of australopithecines and early *Homo* lived at the same time and *all* made tools. However, at the present time no representatives of *Homo* or bifaced tools have been found in the

[1] This view of evolution was usual in Germany and eastern Europe, while the early *sapiens* theory was characteristic of France and England. In general, the nations which accepted Piltdown also believed in very early *sapiens*. In this country, Hooton's views were typically English and Hrdlička's as typical of the area where he was educated. The two theories, supported by strong national biases, originated long before the discovery of the best documented and most important fossils, and before modern archeology. It is necessary to re-examine the primary evidence. At present there is no single fossil of *Homo sapiens* in which a substantial part of the face and of the braincase is preserved which is dated before the last glaciation.

australopithecine sites. The use to which these earliest tools were put is uncertain. What is needed in archeology is more information on their possible use, as this would be of the greatest value in interpreting the adaptation of the animals. At the moment the best guess seems to be that the australopithecines were tool-using gatherers and scavengers.

By around 900 cubic centimeters of cranial capacity, men were making well-shaped, finely formed tools. These men of the middle Pleistocene were certainly hunters, and Spuhler has already mentioned the importance of the hunting habit. Our ancestors can have hunted large animals only with the aid of tools, and the first certain evidence of this comes with the biface or chopper traditions. Even with tools, hunting must have involved cooperation within the group. The use of tools must have led to psychological changes. Compared with their primarily vegetarian ancestors, the men of the Middle Pleistocene must have been more aggressive outside the group and more cooperative inside. Sharing food within the group, cooperating in the hunt, sharing traditions of tool making; these are the positive sides of the changes which came in with large scale hunting. These contrast to the fearsomeness of the hunter to other kinds of animals and lead to the possibility of war. Man has a carnivorous psychology. It is easy to teach people to kill, and it is hard to develop customs which avoid killing. Many human beings enjoy seeing other human beings suffer, or enjoy the killing of animals. After all, public spectacles of beatings and tortures are common in many cultures. This is the characteristic behavior of carnivores and is not seen in apes and monkeys. Much of what we think of as normal human nature may date back only to the hunting traditions of the Middle Pleistocene, while bipedalism and tool use are of much greater antiquity.

TABLE 1

Representative cranial capacities

1200–1500 cc	*Homo sapiens*
900–1100 cc	Java-Peking
450–550 cc	Australopithecines
350–450 cc	Ape (primarily chimpanzee)

A slightly more detailed view of the evolution of cranial capacity is given in table 1. Dr. John M. Roberts, of the University of Nebraska, and I worked on this last year at the Center for Advanced Study in the Behavioral Sciences. This analysis is the result of our numerous discussions at the Center. The first problem in considering the evolution of the brain is to get estimates of size. Since the fossils are not numerous, many are fragmentary, and estimates in the literature vary greatly, we have therefore resorted to a device which we call "representative capacities." These are the capacities which we think best represent a group of fossils. All the most frequently occurring values lie in the range given. All the figures mean is that 450–550 is about as good an estimate of brain size in the australopithecines as 1200–1500 is for the men of the Upper Paleolithic. We have tried to keep body weight approximately the same. For this reason ape capacities are estimated from chimpanzee primarily, rather than including gorilla with three times the body weight. An *Australopithecus*, one might guess from the bones from Swartkrans, might have weighed around 100–140 pounds, and such an animal would not differ greatly in size from chimpanzee or man. We think that there is a real increase in brain size of approximately three time, at comparable body weight, insofar as that can be estimated in the exceedingly fragmentary fossils. Pebble tools go with the australopithecine-size brain; traditions with the Java-Peking primitive-man size brain; and figures for early *sapiens* do not differ from those for Neanderthal. Capacity is a very crude measure of brain evolution and there may well have been numerous other changes which do not show in the fossils. However, increasing the number of cells is very important, as Gerard mentions in chapter 2. Size does reflect increasing number of cells, but due to folding of the cortex it may lead to an underestimate.

In the expansion from an ape-type brain to a human type, the different parts of the brain did not expand equally. Figure 2 is a diagram from Penfield and Rasmussen (1950) showing the way the human body is represented on the cortex. It will be seen in the diagram that the body is very unequally represented in the motor cortex. The areas which are largest are the ones of greatest functional importance. This is true of the mammalian cortex in general. For example, the areas associated with the face are always large. But if the monkey cortex is compared with that of man, it will be found that in the former the area of cortex associated with the foot is approximately as large as that for the hand. When the brain increased in size, the area for hand increased vastly more than that for foot (which remained stable or, perhaps, decreased slightly). This supports the idea that the increase in the size of the brain occurred after the use of tools, and that selection for more skillful tool-using resulted in changes in the proportions of the hand and of the parts of the brain controlling the hand (including primary motor and sense areas, and the areas concerned with the elaboration of skills).

Again, following Penfield and Rasmussen, the areas of the cortex concerned with speech are very large. The reason that a chimpanzee cannot learn to talk is simply that the large amounts of brain necessary for speech are not there. The frontal lobes of man are greatly expanded

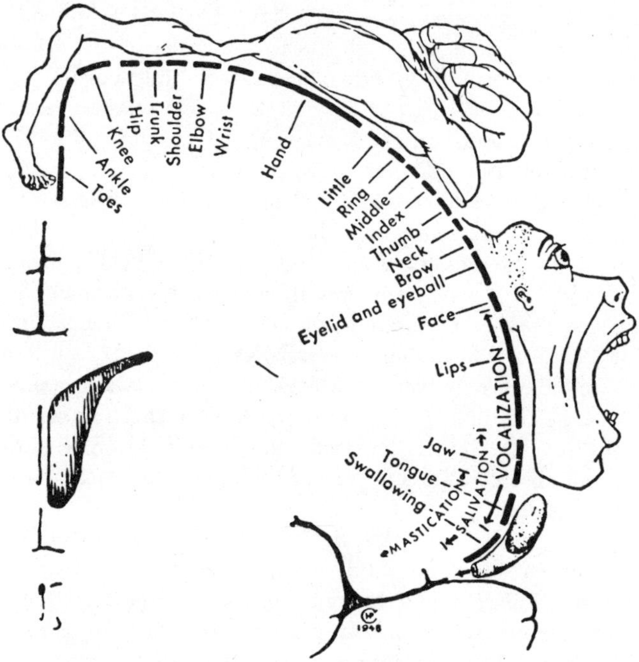

FIG. 2. DIAGRAMATIC REPRESENTATION OF THE PROPORTIONS OF THE MOTOR CORTEX. From Penfield and Rasmussen, *The Cerebral Cortex of Man*, 1950, copied with the permission of the Macmillan Company.

also, and these areas are, at least in part, concerned with elaboration of thought and planning. Foresight and planning are essential to any complicated social life, and in the future it may be possible to demonstrate that the expansion of much of the cortex is directly related to new selection pressures associated with the evolution of complex social systems. Our brains, then, are not just enlarged, but the increase in size is directly related to tool use, speech, and to increased memory and planning. The general pattern of the human brain is very similar to that of ape or monkey. Its uniqueness lies in its large size and in the particular areas which are enlarged. From the immediate point of view, this human brain makes culture possible. But from the long-term evolutionary point of view, it is culture which creates the human brain. Dr. John M. Roberts is investigating the cultural implications of this interpretation of evolution.

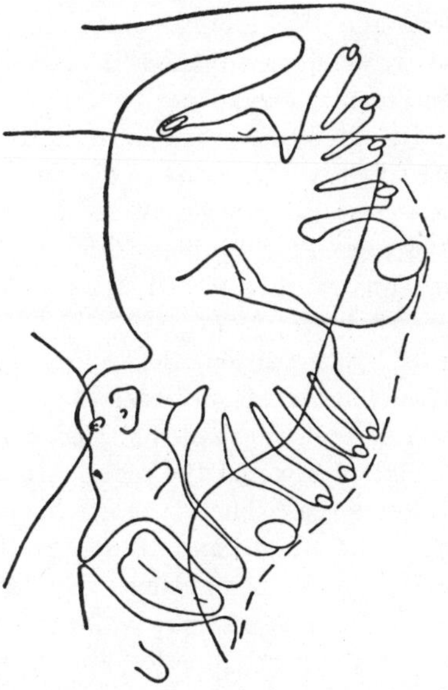

Fig. 3. Motor Homunculus of Monkey. After Woolsey and Settlage. Reproduced from *Handbook of Experimental Psychology,* S. S. Stevens, ed., 1951 by permission of the publisher, John Wiley and Sons, Inc.

Obviously, these speculations range far beyond what can be proved at the present time. But I think that it will be possible to prove much more in the future. In closing, I would like to speculate about these future possibilities. As stated before, a possible evolutionary order is an ape, a tool-using australopithecine, a hunting-speaking complicated tool-making ancient man, and modern man. If we think of such an order, a better understanding of changing diet is of paramount importance in interpreting the fossil record. Spuhler (chap. 1) has already mentioned the lengthening time which the eruption of the teeth take in evolution. The teeth form over a considerable period of time before they erupt, and during this time of formation the developing tooth is affected by its environment. If a growing person is subject to severe starvation or disease, there are characteristic defects built into the structure of the tooth which is being formed at that time. Since the dentition develops over a period of several years, a record of health of most of an animal's youth is frozen into it. The late Dr. E. L. Schuman was carrying out preliminary work along this line, but nobody has developed the necessary methods and systematically applied them. With microscopic x-ray and the tremendous advances in mico-chemical techniques, which Dr. Frederick Thieme is adapting to anthropological uses, major progress seems possible. It seems probable to me that there are records in the teeth and bone which will help us to decide whether animals were primarily hunters or were largely vegetarian. It should be remembered that the teeth of the associated animals can be treated in the same way and may prove equally valuable in reconstructing the ecological conditions.

In the reduction of the dentition, the canines and incisors seem to have decreased in size first. This decrease may have been the consequence of the first use of tools, perhaps of wooden tools which are long since gone. This reduction affected the males primarily, for in female apes the canines are not large. The cause of this sex difference is lost before the australopithecine stage, and it may well be that clubs had already replaced canine teeth in fighting. But among the australopithecines the molar teeth were extremely large. As far as one can tell from the skulls, the temporal muscles were very large too. There is no indication that the chewing part of the dentition started to decrease until after the major tool traditions. This suggests that the preparation of food began in the middle Pleistocene along with hunting of large animals and the biface and chopper tool traditions. Before that time, selection still favored the maintenance of a very large grinding apparatus.

With the reduction of the teeth, (first the canines, then the molars) the face decreased in size. The browridges decreased along with the face, but we have been experimenting on browridges and all the variation cannot be accounted for on the basis of the mechanics of the face. Muscle size and face size account for part of the browridge size, but not for all of it. Browridges are decreased in all domestic mammals.

Pigs, horses, cattle, cats,—all these unrelated forms show characteristic facial changes and no one bred selectively for these features.

The same is true in rats. Tame rats have smaller browridges and faces than wild rats. But here the cause is known. The wild rats have larger adrenal glands, are more prone to rage, and are much more difficult to handle. In choosing animals which are easier to tame, a glandular pattern has been selected which creates differences in the skull. A similar process may have gone on among ancient men, and the most irascible individuals, those who could not fit into the evolving social order, may have become eliminated. This social selection would have affected the form of the human skull.

In summary, it was bipedalism which started man on his separate evolutionary career. But tool use was nearly as early. Biological changes in the hand, brain, and face follow the use of tools, and are due to the new selection pressures which tools created. Tools changed the whole pattern of life bringing in hunting, cooperation, and the necessity for communication and language. Memory, foresight and originality were favored as never before, and the complex social system made possible by tools could only be realized by domesticated individuals. In a very real sense, tools created *Homo sapiens.*

LITERATURE CITED

Hallowell, A. I. 1956 Structural and functional dimensions of a human existence. Quarterly Review Biol., *31* (2) : 88-101.

Oakley, K. 1957 Tools makyth man. Antiquity, *31*: 199-209.

Penfield, W. and T. Rasmussen 1950 Cerebral Cortex of Man. Macmillan Company, New York.

Weidenreich, F. 1947 Facts and speculations concerning the origin of *Homo sapiens.* Amer. Anthrop., *49*: 135-151.

ANIMAL "LANGUAGES" AND HUMAN LANGUAGE[1]

BY CHARLES F. HOCKETT
Cornell University

FROM time to time, anthropologists draw up lists of the kinds of behavior common to all human beings but not shared with other species. One item found on every such list, though under various labels, is the power of speech.

It is true that human language is a remarkably powerful institution. Yet Man is by no means the only animal that carries on communication. Until we can describe in detail just how human language differs from any variety of communicative behavior manifested by non-human or pre-human species, we cannot really know how much or little it means to assert this particular human uniqueness.

So far as I can see, the only way to find out what we want to know in this area is as follows. First, we must itemize those design-features of human language which seem to be of crucial importance in making it possible for language to do what it does. Second, we must examine all the non-human systems of communication that can be found, to see to what extent each crucial design-feature of human language recurs. It goes almost without saying that in carrying out this program we must not allow ourselves to neglect any point because of its apparent triviality. Indeed, the most important items will surely tend to be hidden from our view because they are right under our noses.

My own research on this problem has so far yielded a list of seven properties of human language that seem to merit consideration. None of these is my own discovery; my role has been merely that of bringing the seven together because of their possible relevance for the particular problem, and, in some cases, of rephrasing a point in such a way as to render it easier to handle.

The first of the seven may be called *duality of patterning*. For some

[1] This is an interim report on a project begun during 1955-6 at the Center for Advanced Study in the Behavioral Sciences. Research continues as time allows, and a fuller discussion will eventually appear.

Although the practice is virtually unprecedented with works smaller than whole books, I should like to dedicate this article to Alfred L. Kroeber, in appreciation of his stimulation and encouragement.

reason, this seems to be the most difficult to explain. I shall therefore begin with a simple nonlinguistic example.[2] Suppose that Paul Revere and his colleague had needed a total repertory of several hundred messages, instead of just two. It would have been inconvenient to have had several hundred lanterns on hand in the Church tower. But it could have been agreed that each message would take the form of a row of 5 lights, each one either red, or yellow, or blue. Then only 15 lanterns would have been needed—one of each color for each position —but the system would have provided for a total of $3^5 = 243$ different messages. We assume that meanings would have been assigned only to the whole messages, so that, for example, "red light in first position" would not have had any separate meaning of its own, but would merely have served to distinguish certain messages from others. This modified Paul Revere system would then show what we mean by duality of patterning: a set of conventions in terms of *smallest meaningful elements* —here the whole messages—and also a set of conventions in terms of *minimum meaningless but differentiating ingredients*—the three colors and 5 positions.

Any human language has this same design-feature. In English, for example, the phoneme /b/ at the beginning of an utterance has no meaning of its own, but merely serves to distinguish *beat* from *meat*, *bat* from *rat*, and so on. Phonemes are the minimum meaningless but differentiating ingredients in a language; the smallest meaningful elements are what we call morphemes.

The principle of duality is a great convenience in any communicative system where the repertory of messages must be large, since a relatively small stock of minimum meaningless ingredients can be ordered into a very large stock of combinations to which meanings can be assigned. Many human communicative systems, though not all, show duality, perhaps just because of its economy. But wherever it turns up, it seems to have been developed by analogical transfer from language. No animal communicative system on which I have information incorporates this principle. In duality, then, we may have found one genuine uniqueness of human behavior.

The second key property may be called *productivity*.[3] It is a

[2] I have published a fuller description of duality, using the same example, (Hockett, 1955). See also Martinet (1949), where the hints implicit in linguistic discussions since the Middle Ages are explicitly spelled out for the time.

[3] According to Wells (1949) this vital property of language was first overtly pointed out by the philosopher Bertrand Russell.

commonplace that a speaker may say something that he has never before said or heard, and be understood without either speaker or audience being aware of the novelty. That is, we coin a new utterance on the analogy of familiar ones. We build a new utterance out of parts that have occurred in previous utterances, putting them together by patterns familiar from previous utterances. Since the audience, as well as the speaker, has had previous experience with the parts and patterns, the new combination is understood.

Gibbons, reported to be the noisiest of our near kin, seem not to do this (Carpenter, 1940). They have a repertory of some dozen different calls, but no matter how novel may be the situation in which a gibbon finds himself, he is constrained to respond to it with one or another of this small finite set: he does not coin a new call by putting parts of old ones together.[4]

But productivity is not unique to human beings. It recurs in bee dancing (von Frisch, 1950). A worker can report a source of nectar at a location which neither she nor her coworkers has ever visited before. The mechanisms involved are quite different, but at the moment this is irrelevant.

The third property is *arbitrariness.*[5] A message means what it does *iconically* if it resembles its meaning in physical contours, or if the whole repertory of possible messages shows a geometrical similarity to the whole repertory of possible meanings. Otherwise the semantic relationship is arbitrary. Human language is almost wholly arbitrary. For example, there is no similarity between the sound of the word *dog* and the sight, sound, or smell of a dog. Nor is the difference between the sounds of the words *dog* and *cat* in any way similar to the difference between dogs and cats. In this framework, onomatopoetic forms constitute only faint traces of iconicity.

By contrast, bee-dancing is iconic. The rate of the dancing is inversely proportional to the distance from hive to source of nectar,

[4] A gibbon may vary the intensity and duration of a single call. This, however, finds its human counterpart not within language but in the nonlinguistic (though communicatively relevant) variations of volume, pitch, tempo, and tone-quality which accompany the linguistic ingredient in speaking. See Trager, (1958).

[5] Arbitrariness and iconicity have been the source of more trouble than any other aspect of communicative behavior, from the medieval dispute between "realists" and "nominalists" down to the present. Yet Charles S. Pierce made the distinction quite clearly in the nineteenth century; and Saussure (1916) took the arbitrariness of linguistic symbols as one of the cornerstones of his work.

and the angle between the line of dance and the vertical is directly proportional to the angle between the line of sight to the sun and that to the source of nectar.

The contrast between arbitrary and iconic is also exemplified by digital and analog computers. An analog computer is often beautifully adapted for a narrow function and worthless for anything else. Just so, bees can talk about nectar and hive-sites; human beings can talk about anything.

Yet arbitrariness is not a human prerogative. It characterizes gibbon calls, and probably most other vocal systems among mammals and birds (King, 1955).

The fourth property is *interchangeability.*[6] Any speaker of a language is theoretically capable of saying anything that he can understand when someone else says it. Actual limitations on this theoretical interchangeability are practical or accidental, imposed by institutionalizations of social roles or by pathologies of individual capacity. So likewise, it would seem, for gibbon calls and for bee dancing. But in some systems there is not even theoretical interchangeability. In the signalling of certain fish, male and female sticklebacks, during courtship, it is inherently impossible for roles to be exchanged (Tinbergen, 1953).

The fifth is *specialization.* An act cannot be communicative unless it involves *triggering* (Bloomfield, 1942). But any act, merely as physical activity, also has direct *energetic* consequences. When these two kinds of consequences are unrelated, we say that the communicative system is specialized. When sticklebacks are courting, an essential ingredient in one of the signals from the female is her seasonal appearance, with abdomen distended by roe. The direct consequences of this distension are obviously closely related to the trigger consequences in the male's behavior, and so the system is unspecialized. When Mrs. Jones calls out "Dinner is ready," her family are triggered into coming to the table; but the direct consequences of her action are only some minor flurries in the air, damping out with a tiny rise in the temperature of air and walls. Here we have a high degree of specialization.

The sixth property is *displacement.* We can speak about sticklebacks, or about shoes and ships and sealing wax, when none of those things

[6] So far as I know, this is the first overt discussion of interchangeability and of specialization (the next property to be treated). These properties of human language come into view only when one examines certain types of animal interaction which lack them—types of interaction which many scholars would refuse to subsume as "communication" at all.

are around. Our speech can be removed in space and time from what we are talking about. Bee-dancing is also displaced: the worker goes back to the hive to give her report. Gibbon calls are reported not to be displaced. If a gibbon finds food, he stands by it and emits the proper call to the rest of the band.

The seventh property is *cultural transmission.* A behavior pattern is transmitted culturally if it is not only learned but *taught,* and if the teaching behavior, whatever it may be, is also learned rather than genetically determined.[7] Human language is transmitted in this way, as are many other facets of human life. There is evidence for culture of a rather thin sort among the hominoid apes, and even stronger evidence in the case of waterfowl (Hochbaum, 1955), but it is not clear in either case that it is the *communicative* systems of the species that are so transmitted. Thus it may be that gibbon calls are passed down genetically, even if in some other connections gibbons, like chimpanzees, show traces of culture. For bees, we can be almost certain that genetics rather than culture is involved.

There are other universal properties of human language that might be considered, and there are certainly a great many animal communicative systems about which we are as yet in total ignorance. However, I am now going to venture a historical hypothesis based on the partial findings, fully cognizant that subsequent research may upset it.

The proto-hominoids, common ancestors of the hominoid apes and ourselves, lived some ten to fifteen million years ago.[8] It seems reasonable to suppose that they had a vocal-auditory communicative system something like that of present-day gibbons, characterized by specialization, interchangeability, and arbitrariness, properties which we may suspect were inherited from early mammalian times. It would also seem reasonable to suppose that they had a small amount of cultural transmission, though perhaps not in the domain of their most highly developed communicative system.

This proto-hominoid system was the precursor of human language. In order for it to evolve into genuine language, it had to develop four new properties. Of these, I believe the first to develop was productivity,

[7] Note that if we drop the requirement that the teaching behavior itself be learned, the transmission may be what the geneticists call a "maternal effect," and thus basically genetic.

[8] Based on the antiquity of *Proconsul,* who may have been our remote uncle or up-generation cousin rather than direct ancestor, but is nevertheless the nearest to a proto-hominoid so far discovered.

because it is easier to explain the inception of the other three if productivity came first, and not impossible to account for the development of productivity in the absence of the other three.

An unproductive system, with a finite repertory of signals, can be rendered productive only by one mechanism with which we are familiar: a mechanism known to linguists as *blending*. In blending, a speaker confronts a situation partly like two different types of situations previously experienced. He starts to say two words at once, and comes out with something different from either but partly similar to both. An example is Lewis Carroll's *slithy*, based on *lithe* and *slimy*. Conscious planning, however, is not necessary: I once hesitated between "Don't shout so loud" and "Don't yell so loud," and actually said "Don't shell so loud." A number of English words are known or suspected to have originated in this way.

I imagine that in pre-human times our remote ancestors would occasionally produce such blends, perhaps under the stress of unusual circumstances, despite the fact that their repertory of vocal signals was basically closed. Probably in thousands of cases no one responded appropriately, the blend being too radical an innovation to be intelligible. But sooner or later a few blends were communicatively successful. As soon as this happened, the closed circle was broken, and productivity was on its way.[9]

The conventions of the earlier unproductive system could be transmitted genetically, and perhaps were. But productivity, conjoined to arbitrariness, implied that members of the species were acquiring new habits from each other. It seems to me that this must automatically have called into play any capacity for culture our ancestors of the time had. Also, the more flexible communicative system had considerable survival value, and bestowed survival value on the capacity for culture (and thus perhaps also, indirectly, on increased cranial capacity).

Given arbitrariness, productivity, and cultural transmission, the young had to learn the conventions of the system before they could use it. As soon as the young were sometimes taught a habit out of the exact context in which the response would normally be evoked, circumstances were ripe for the development, or sharp increase, of displacement. The innate capacity to learn to *imitate*, outstanding in hominoids though most fully developed in Man, must have helped here.

[9] An innovation consisting of two old signals in immediate juxtaposition counts, in the case of a closed system, as one kind of blend.

Finally, increasing productivity led to the development of duality of patterning. The exact causal connection is not clear, but I shall speculate. As the once closed system came to have a larger number of minimum meaningful elements, the latter came to be more and more similar to each other in physical contours. In time an impasse was reached. Further increase in their number would surpass the discriminatory powers of the speakers, if they had to keep the signals apart by listening to them as wholes. But this impasse could be avoided by paying attention to the smaller constituent features of sound occurrent in the wholes. It is only after the fact, of course, that we can know that the second alternative was followed. This line of reasoning derives, rather dangerously perhaps, from an observation of the language-learning process of contemporary children. Some children acquire rather large vocabularies, with excellent pronunciation; but then suddenly seem to discover that words are kept apart by constituent sounds rather than holistically; their pronunciation then deteriorates for a period, and the rate of vocabulary-learning is slowed, until they build up the necessary skills for producing and distinguishing phonemes.[10]

These successive evolutionary changes, leading to genuine language, presumably did not begin more than 10 to 15 million years ago, since our nearest non-human cousins do not show the consequences; they may have begun much earlier. They were concluded at least 50,000 years ago, and may have been completed much earlier. This second date is based on a rough estimate of the time which would have been required for all the languages of the world today to have differentiated from a single parent language, on the assumption that they are all related. I do not recommend this assumption, which is highly dubious; but it affords us our only way of directly estimating a *terminus ad quem.* Indirect inferences, based on archeological reconstructions of paleolithic life, would suggest a much earlier terminal date. Quite possibly *Pithecanthropus,* if not *Australopithecus,* shared with *Homo* the power of speech.

[10] Again, in the history of the Chinese writing system it early became necessary, for mnemonic reasons, for new characters to be partly similar in visual shape to old ones. Here, however, full-fledged duality was never reached.

LITERATURE CITED

BLOOMFIELD, LEONARD 1942 Philosophical aspects of language, in Studies in the History of Culture, George Banta Publishing Company, Menasha, pp. 173-177.

——— 1933 Language, Henry Holt, New York.

CARPENTER, C. R. 1940 A field study of the behavior and social relations of the gibbon, Comparative Psychology Monographs, *16*, No. 5.

HOCHBAUM, H. ALBERT 1955 Travels and Traditions of Waterfowl, University of Minnesota Press, Minneapolis.

HOCKETT, CHARLES F. 1955 "How to Learn Martian," Astounding Science Fiction, *55* (May) : 97-106. Reprinted in GREENBERG, MARTIN, ed., Coming Attractions, New York, 1957.

KING, JOHN A. 1955 Social behavior, social organization, and population dynamics in a black-tailed prairiedog town in the Black Hills of South Dakota, University of Michigan Contributions from the Laboratory of Vertebrate Biology, No. 67.

MARTINET, ANDRÉ 1949 La double articulation linguistique, Traveaux du Cercle Linguistique de Copenhague, *5*: 30-37.

SAUSSURE, F. DE 1916 Cours de Linguistique Générale, edited posthumously by CHARLES BALLY and ALBERT SECHEHAYE, Payot, Paris.

TINBERGEN, N. 1953 Social Behavior in Animals, Methuen, London, and John Wiley, New York.

TRAGER, GEORGE L. 1958 Paralanguage: a first approximation, Studies in Linguistics, *13*: 1-12.

VON FRISCH, KARL 1950 Bees, their Vision, Chemical Senses, and Language, Cornell University Press, Ithaca, New York.

WELLS, RULON S. 1949 Language, *25*: 323.

BASIC SOCIAL CAPACITY OF PRIMATES [1]

BY HARRY F. HARLOW

University of Wisconsin

THE existence or nonexistence of cultural patterns in subhuman primates may best be determined by observation and comparison of the behaviors of primate groups in their native habitats. Fortunately, a considerable body of such data already exists, as exemplified by the detailed and insightful studies of primates in the wild, including the observations of Carpenter (1934) on the howler monkey, Nolte (1955) on the bonnet macaque, Zuckerman (1932) on the baboon, and last, but far from least, the magnificent unpublished studies by Washburn on three different baboon cultures.

Beyond, but not above, such data there exist laboratory studies, both experimental and observational, on more limited aspects of primate behaviors, the kinds of specific traits and capacities which it is reasonable to assume must exist if cultural acquisitions are to be passed down from generation to generation in the wild. In this regard, we may be interested in both the kind of capacities making possible existing human cultures and the kind of developing capabilities that may have gradually led to the formation of our own cultural environment.

Culture is dependent upon the ability of an animal within a social group to acquire from another group member information previously learned by the latter member. To obtain such information it is essential that the animal, particularly the young animal, observe other members of its social group and duplicate their behaviors either explicitly or implicitly.

It has long been known that monkeys in a relatively free situation do observe other monkeys and consistently duplicate their behaviors. The early literature was summarized by Watson in 1914, and we quote limited sections: "With Kinnaman we found that if one monkey discovered a hole and peeked into it, another would generally push him aside and peek into it in his turn. This was observed many times. More recently we confirmed these observations many times in the case of a

[1] The author's research reported at the end of this paper was supported in part by Grant M-772, National Institutes of Health, and in part by funds received from the Graduate School of the University of Wisconsin.

mother and baby. One day the baby left the mother's arms and peered under a sill at a brass lock which could be seen from his viewpoint but not from hers. He pushed his arm through and attempted to pull the lock forward. The mother left off eating, came and adopted the same position and reached her arm through and attempted to grasp the lock. The two acts were identical in technique" (pp. 284-285).

Anyone who has attempted to tame and adapt groups of monkeys to the experimenter or to an apparatus knows that the behavior of the tamer subject greatly influences the behavior of those less tame. We again quote representative observations from Watson: "Two of the monkeys were quite friendly with each other. J was an adult monkey and B a young one. In a short time an attachment sprang up between these monkeys which persisted for several years. . . . When J went to one part of the cage, B followed. If, while sitting on the shelf, a pan of water or bowl of milk was placed upon the ground, B would not come down to drink if J did not precede him. J formed the habit of jumping on to the experimenter's shoulder when he entered the cage and called. B formed the same habit, but if J for any reason refused to come, B refused also" (p. 283).

Dr. Mason, at the Wisconsin Primate Laboratory, has made extensive and detailed observations under controlled conditions of the behavior of infant monkeys separated from their mothers at birth and raised subsequently in individual cages. Within the first month of life there developed in these infants strong tendencies to observe the exploratory-manipulatory behaviors of the other member of the test pair, and subsequently to engage in and esentially duplicate these activities. This imitative responsiveness becomes stronger with age, clear-cut individual and pair differences develop, and individual imitative responsiveness becomes influenced by peer status. Dominant animals both observe and duplicate the behavior of the subordinate, whereas subordinate animals tend to observe, but not duplicate, the behavior of the dominant member of the pair, or to defer duplication of this behavior until the dominant animal has left the area.

In Watson's 1914 review he concluded that there was no evidence in primates of imitative learning involving more complex tests, such as problem-box devices and tools. However, Warden and Jackson in 1935 tested 15 rhesus monkeys in the Warden duplicate-cage imitation apparatus and obtained positive evidence of imitative learning on problem-box devices, ranging in difficulty from pulling a chain to the operation of two latches. The opportunity for imitative learning was provided by illuminating the cage of the trained monkey, which then solved the particular problem five times, usually within a 30-second period. The light was then turned on in the imitator's cage, and it was allowed 60 seconds to solve the observed problem. Six such opportunities were provided on each of four tasks. Successful imitation occurred on 46% of the tests, and eliminating 6 subjects that did not adapt to the situation raised the percentage of successful imitation trials to 72. The most successful animal imitated in 23 of the 24 tests presented. Subsequently, Warden, Fjeld, and Koch (1940) conducted similar tests on cebus monkeys and obtained similar results.

There are many formal psychological experiments in which the animal can solve the problem only by observing the behavior of the experimenter. This is true, for example, in the delayed-response test run by the direct method. To solve the problem, the animal must observe the foodwell or cup in which the experimenter places the food before covering it and the unbaited well with the identical stimuli, thus initiating the delay period. Primates readily adapt to such procedures, and so, for that matter, do other animals, including cats, dogs, raccoons, and even rats. Harlow (1944) trained a monkey to solve a series of discrimination problems when the only cue to the correct stimulus was the fact that it had been tapped by the experimenter. If primates can solve formal laboratory problems when the only clues to solution are those obtained from observing the experimenter, it seems only reasonable to assume that they can solve such problems when the cues are obtained by observing or imitating a fellow member of their species.

The first indication that subhuman primates might solve a discrimination problem merely by watching the solution of the same problem by another animal was obtained by Crawford and Spence (1939), who used chimpanzees as subjects. Perfect learning in this situation was obtained in one of 11 subject pairings, and partial learning was found in two other cases before the imitator showed clear behavioral indications of loss of interest.

Miller and Murphy (1956) have recently compared the learning of discrimination and oddity problems by rhesus monkeys tested individually and in pairs, and they found that social interaction improved performance significantly on a series of oddity problems and significantly improved the performance of previous discrimination tasks. Data from a study by Darby and Riopelle (1958) give even more striking proof of the ability of primates to learn discrimination problems entirely on the basis of observation of a partner. They used an adaptation of the

Crawford-Spence technique in which stimulus objects were placed on a tray between two opposed animal cages and the subjects could observe each other and the test situation. A total of 1000 problems was given, on half of which each animal served as demonstrator and on the other half as observer. Each problem lasted only four trials, one free-guess trial by the demonstrator and three test-trials by the observer. The imitators became progressively more proficient in solving problems after a single observational trial. Thus, the imitator learned to solve successive problems with progressively greater facility, *i.e.*, to make very significant progress toward the formation of a discrimination learning set. In view of the fact that the formation of discrimination learning sets is by no means a simple task for the subhuman primate, a truly amazing capability of imitative learning was demonstrated. Indeed, not only could the observer select the same object that was chosen by the demonstrator when it covered food but the animal could also take into account errors made by the demonstrator.

There is reason to believe that there are subtle factors influencing imitative learning by primates, human or subhuman. It is obvious that the nature of the social interactions already existing between members of the test pair is of primary importance. It is more than possible that the conditions existing in nature have been more favorable in this regard for producing imitative learning than have the conditions in the laboratory. Mother monkeys, for example, may have far more insight into the importance of these variables than do experimental psychologists. Being now able to identify some of these variables, we hope that both more spectacular and more illuminating experimental data will be provided by comparative psychology laboratories in the future.

Although the existent experimental data give us no evidence that anthropoid apes have greater capacity to profit from imitative learning than do Old World monkeys, there is enough observational evidence to evoke at least mild skepticism about this point. Köhler's (1925) beautiful pictures showing intent observation by single, or even multiple, chimpanzees while a single member of the clan works on the solution of a multiple-tool problem, have never been duplicated with monkeys. It is more than possible that the chimpanzee's capabilities along these lines have been even less thoroughly exploited than have the monkey's. An extremely interesting set of observations on imitative behavior in a home-reared chimpanzee was made by Hayes and Hayes (1957) on their adopted chimpanzee, Vicki. From the age of 17 to 34 months, Vicki was given occasional training on a series of tasks intended to establish a learning set for learning by imitation. They rewarded Vicki for imitating actions accompanied by the command, "Do this," training her to imitate actions that had no intrinsic play value. A list of 70 items, ranging from simple tasks such as hand clapping and head patting, to more difficult problems, including a three-lever problem box and a key-and-padlock task. The Hayeses were in general quite successful in this effort, and there can be no question but that a set was established in Vicki to attempt imitation upon request. The attempts were made in many cases after delay and in some instances even under conditions of subterfuge. The attempted imitation in no way implied successful imitation, particularly for difficult problems. Some failures were attributable to motor limitations, and some were doubtless dependent upon learning limitations. But the heart of the problem was not that of teaching the chimpanzee to solve all problems by imitation, but to teach the subject to try to imitate, and in this the Hayeses were most successful.

The existence of a human-type culture is dependent upon learning capability and the capacity to transmit this learning to subsequent generations. Without language the capacity to transmit previous learning is dependent upon either such forms of imitative learning as we have described, or a simpler type of learning in which the infant follows the group, and the movement and behavior of the group are such as to place the infant repetitively in the situations where the infant learns by direct experience the danger responses or affectional responses essential to individual and group survival.

Laboratory researches of the last quarter-century have demonstrated that both chimpanzees and rhesus monkeys (and presumably other apes and other monkeys) possess a heretofore unsuspected ability to master the general kinds of problems that seem to characterize human intelligent behavior. It is difficult to describe these problems briefly and in a nontechnical manner, but we may indicate their complexity by pointing out that many of them, such as the complex oddity and matching problems, are similar in nature to test items in standard intelligence tests for children, and in diagnostic tests for brain-damaged adults. Others of these tests, including the development of rather abstract concepts such as triangularity, color, and number, are tasks which may prove troublesome for the preschool or primary school child. All data we have to date seem to indicate that the subhuman primates learn such problems with a facility quite beyond that of any other subhuman mammals or other classes of animals.

Furthermore, there appears to be another kind of learning ability essential to any kind of complex learning or thinking which is, or approaches, being the private property of the primate order. This is the ability to effectively transfer the learning of a particular problem to the learning of another problem of the same kind or class, a phenomenon described as the formation of learning sets. If, for example, we train a macaque monkey for 6 trials on each of a series of 100 discrimination problems, the monkey makes progressive improvement on these problems and eventually solves any of the problems with perfect insight. This kind of problem can apparently be solved by all monkeys and by most human beings. A study by House and Zeaman (1958) indicates that it is beyond the capacity of human imbeciles, and similar studies have obtained camparable results with other retarded people.

We have no proof that the ability to learn abstract problems of this type has contributed to the kinds of subhuman cultures so admirably described by Washburn in his field study of baboons, but the ability to learn multiple and even more complex problems of the kind described is an essential prerequisite to the evolution of our own human cultures.

There has been speculation that one of the factors that has contributed to human culture is the long period of infancy or dependency and the ability of the human infant to learn during this period of time. The anthropoid apes and the monkeys also undergo relatively long periods of infancy or dependency, and it is now quite clear that they have developed relatively great capacity to learn during this time. If we assume that the human being is dependent until attaining or approaching adolescence, we may note that this is the time at which learning capability approaches a maximum, and the human animal has been an effective learner for approximately ten years.

We have no definitive data on the period of dependency of the macaque (or other) monkey, but observational evidence suggests a period of two or three years, and we may be approaching complementary data from experiments at the Wisconsin Primate Laboratory. Though this period is short in comparison with that of man, the macaque becomes an effective learning machine much more rapidly than does the child. The macaque can learn conditioned paid responses in the first two or three days of life, simple spatial reflexes by 15 days of age, black-white discriminations by 15 days of age, and learning sets by the end of the first year. Thus, the monkey, like the human being, is a capable and efficient learning animal during this period of developmental dependency. Equally precise data do not exist for the chimpanzee (or any other anthropoid), but its position both in terms of period of dependency and maturation of learning capacities is without doubt intermediate between that of monkey and man. This interrelationship between dependency and abstract learning capabilities has failed to give either monkeys or apes a human-type culture, but it may well be a function that has proved important in the evolution of man's capacity for culture.

The evolution of culture, particularly human-type cultures, which involve complex interrelations among large groups of people, is only possible if there exist positive cohesive forces stemming from innate affectional factors which may be modified, generalized, and enhanced by learning. The existence and operation of these affectional factors has been described and assigned proper importance by Carpenter, Zuckerman, and Washburn for the howler monkey, macaque, and baboon, and by Kohler (1925) and the Yerkes (1929) for the chimpanzee. These affectional responses are above and beyond those associated with sexual behavior, and important as these latter behaviors may be, I believe we may safely assume that sex alone is not enough to account for the cohesive forces essential to the development of human cultures.

The description and enumeration of the multiple patterns of affectional responses of monkeys and apes can be accomplished more efficiently and comprehensively by the study of the free-ranging primates than by the study of the primate as a laboratory animal. The contribution which the laboratory may make, above and beyond naturalistic study, is the analysis of the nature and, to a limited degree, the development of these affectional forces. At the Wisconsin Primate Laboratory we have initiated a series of studies designed to determine the various innate affectional factors and to determine the importance and role of these separable factors. We do this by separating the infant from the mother at birth, and then raising these orphans on artificial and inanimate pseudomothers whose stimulus qualities—such as body texture, presence or absence of lactation, motility, vocalization, and presence or absence of appendages—can be controlled and separably manipulated.

In plate I, top, we show two of these pseudomothers described arbitrarily as good and bad pseudomothers. The good pseudomother is made of a block of wood covered by sponge rubber which is sheathed by a terry-cloth skin. The bad pseudomother is made of hardware cloth, a substance entirely adequate to provide support and nursing capability. In one of our test situations, the infant is free to leave its living cage at any time and contact either pseudomother. For half the infants the bad pseudomother lactates and the good pseudomother does not; for the other infants this condition is reversed. In figure 1 we present data showing the total time spent on the cloth and wire pseudomothers under the two conditions of feeding. These data make it obvious that body texture is a variable of overwhelming importance in the development of affectional responses, and lactation is a variable of negligible importance whether

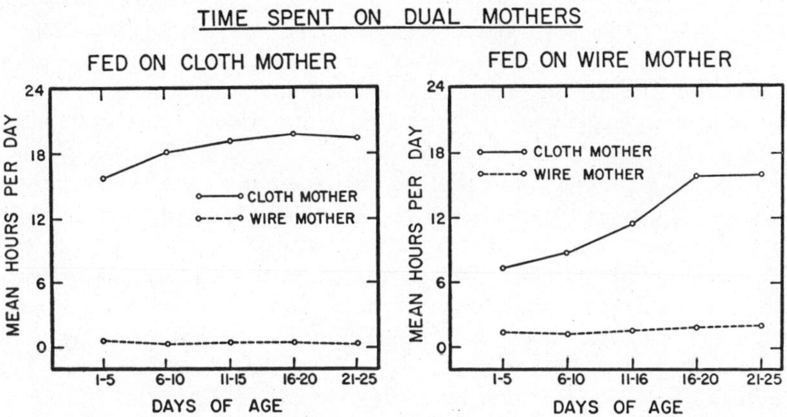

Fig. 1. Time Spent on Dual Mothers.

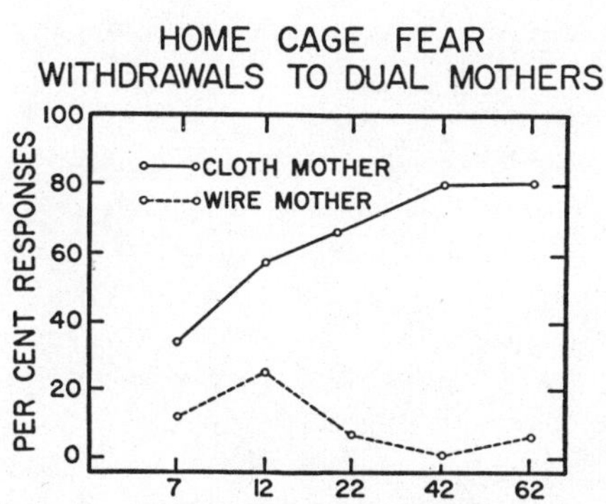

Fig. 2. Home Cage Fear Withdrawals to Dual Mothers.

this relates to innate responsiveness to the pseudobreast or to any learned or derived motives built upon the response of nursing. Indeed, it is possible that the only function of nursing as an affectional variable is that of insuring frequent body contact. Body surface is, of course, a variable with far greater potentialities for primary stimulus generalization than is a lactating breast, since all monkeys at all times have a body surface, and no monkey at all times has a lactating breast.

One function of the mother or mother surrogate is to provide a haven of safety for the infant in times of fear or danger. The frightened or ailing child clings to its mother, not its father, and this selective responsiveness in times of distress, disturbance, or danger may be used as a measure of the strength of affectional bonds. We have tested this kind of differential responsiveness by presenting our infants with various fear-producing stimuli such as this moving toy teddy bear shown in plate I, bottom. The data on differential responsiveness are presented in figure 2, and it is apparent that the cloth pseudomother is highly preferred over the wire one, and this differential selectivity is enhanced by age and experience. The variable of nursing appears to be a variable of absolutely no importance in this situation as indicated by our relatively scanty data at the present time. Contact comfort is one skin game which is essentially good and not evil or improper. If the infant orphans are raised on a single cloth mother they invariably seek her succor in times of terror. If the infant orphans are raised on a nursing, wire mother, they never seek her succor in times of fear, but, instead, throw themselves prostrate on the floor, wrap their head in their arms, and scream in terror.

Pseudomother-raised monkeys, and control monkeys of comparable ages, raised for 14 days on a cheesecloth diaper and henceforth in a bare wire cage, were introduced into the strange environment of a room measuring 6 feet by 6 feet by 6 feet and containing multiple stimuli known to elicit curiosity-manipulatory responses in baby monkeys. All the subjects were placed in this situation twice a week for 8 weeks with no pseudomother present during one of the weekly sessions and the appropriate pseudomother, the kind which the experimental infant had always known, present during the other. A cloth diaper was always available to all monkeys during all sessions.

The animals raised on the good cloth pseudomother always rushed to the mother when she was present and clutched her with loving endearment. Then without exception, they began exploring the room and manipulating the stimuli, and some even brought the stimuli to or around the mother, as if they wished her to share their treasures. The behavior of the infant raised on the good pseudomothers was quite different when she was absent from the room. Emotional indices such as vocalization, crouching, rocking, and sucking increased sharply, and many of the animals would rush to the center of the room where the

PLATE I.

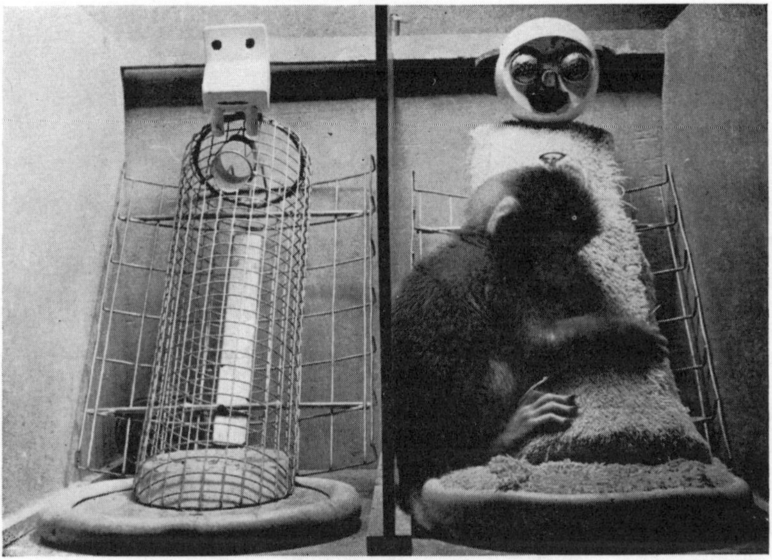

Wire and Cloth Pseudomothers.

Representative Fear-producing Stimulus.

PLATE II.

Contact with Pseudomother in Strange Environment.

Monkey Responding in Modified Butler Box.

mother was always positioned, and then run rapidly from object to object, screaming and crying all the while. Continuous, frantic clutching of their own bodies was very common, even when they were not in the crouching position. These monkeys frequently contacted and clutched the cloth diaper, but this action never brought surcease to their sorrow.

The behavior of the control monkeys and those raised on the wire pseudomothers was similar under all test conditions. They simply froze in the strange situation and made no attempt to contact the wire pseudomother or the diaper or any test object during any part of any test session. The experimental and control infants ranged in age from 7 to 108 days, and there was no responsiveness to the diaper or the wire

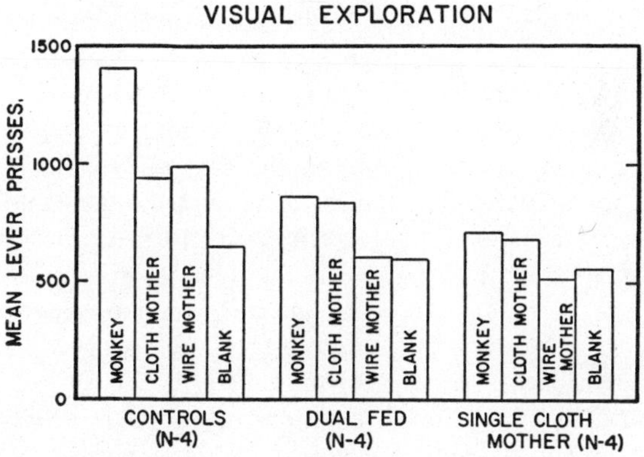

FIG. 3. DIFFERENTIAL VISUAL-EXPLORATION RESPONSIVENESS.

mother which could increase or decrease. These data suggest that man cannot live by milk alone.

A number of years ago Butler (1953, 1954) demonstrated that monkeys enclosed in a box would open and reopen a door for hours on end, for no other motivation than of viewing. Furthermore, he showed selectivity in rate and frequency of door opening to stimuli of differential attractiveness. We have utilized this apparatus to measure in a positive manner the strength of affectional responsiveness in our infants, as illustrated in plate II, right. The data of figure 3 show equally strong responding to the cloth pseudomother and another infant monkey and no greater responsiveness to a wire pseudomother than to a blank cage. These data are again independent of the kind of mother which nursed the infant.

The fact that the comfort of contact is the variable of prime importance in the development of filial affection is no more surprising than the finding that good pseudomothers, as inanimate as if they were made from a block of wood, which they were, elicit such strong and persistent affectional responses. The infant seeks them when theatened, feels secure in their presence, views them with loving adoration, learns to solve puzzle boxes and discrimination problems to attain them, and grieves in their absence. Our oldest pseudomother-raised monkeys are now 6 months of age, and their responsiveness to their mothers seems to be increasing rather than decreasing. We are now beginning studies to determine how strong and how persistent are the affectional responses when the infant and pseudomother are separated. Our data are limited, but we are certain that affectional responses once formed are very persistent, perhaps as persistent as responses conditioned to pain. It appears that the affectional responses to the pseudomothers generalize adequately to real monkeys, although we have not yet subjected this problem to experimental analysis. We are beginning studies that will enable us to compare the strength of affectional responses to pseudomothers and restrained real mothers, lactating and nonlactating, and to restrained real fathers.

Anthropologists and evolutionists have emphasized the importance of freeing the forelimbs from locomotor functions for activities of exploration, tool construction, and general language. But another important function of arms is their capability of affording intimate and maximal contact comfort to the infant, and it is possible that this manual maximizing of contact comfort for long periods of time, time during which the infant has developed excellent learning capabilities, has contributed to the strong and enduring individual affectional attachments important to the development of any complex culture.

Our experiments on the nature and development of affectional responses in rhesus monkeys are not designed to attack directly the problem of the evolution of culture within the primate order, but it is our hope and faith that they provide very useful information concerning a kind of response of vast importance to the social behavior of all animals in all societies and cultures.

LITERATURE CITED

BUTLER, R. A. 1953 Discrimination learning by rhesus monkeys to visual-exploration motivation. J. comp. physiol. Psychol. *46*: 95-98.

BUTLER, R. A. AND H. F. HARLOW 1954 Persistence of visual exploration in monkeys. J. comp. physiol. Psychol. *47*: 258-263.

CARPENTER, C. R. 1934 A field study of the behavior and social relations of howling monkeys. Comp. Psychol. Monogr. *10*, No. 2: 1-168.

CRAWFORD, M. P. AND K. W. SPENCE 1939 Observational learning of discrimination problems by chimpanzees. J. comp. physiol. Psychol. *27*: 133-147.

DARBY, C. L. AND A. J. RIOPELLE in press Observational learning in the rhesus monkey. J. comp. physiol. Psychol.

HARLOW, H. F. 1944 Studies in discrimination learning by monkeys: II. Discrimination learning without primary reinforcement. J. genetic Psychol. *30*: 13-21.

HAYES, CATHY 1957 The Ape in our House. Harper & Bros., New York.

HOUSE, BETTY S. AND D. ZEAMAN 1958 Reward and non-reward in the discrimination learning of imbeciles. J. comp. physiol. Psychol., *51*: 614-618.

KÖHLER, W. 1925 The mentality of apes. Harcourt, Brace and Co., New York.

MILLER, R. E. AND J. V. MURPHY 1956 Social interactions of rhesus monkeys: II. Effects of social interaction on the learning of discrimination tasks. J. comp. physiol. Psychol. *49*: 207-211.

NOLTE, ANGELA F. 1955 Observations on the behavior of free ranging *Macaca radiata* in southern India. *Zeitschrift für tierpsychologie, 12*: 77-87.

WARDEN, C. J., H. A. FJELD AND A. M. KOCH 1940 Imitative behavior in cebus and rhesus monkeys. J. genetic Psychol. *56*: 297-310.

WARDEN, C. J. AND T. A. JACKSON 1935 Imitative behavior in the rhesus monkey. J. genetic Psychol. *46*: 103-125.

WATSON, J. B. 1908 Imitation in monkeys. Psychol. Bull. *5*: 169-178.

YERKES, R. M. AND A. W. YERKES 1929 The Great Apes. Yale Univ. Press, New Haven.

ZUCKERMAN, S. 1932 The Social Life of Monkeys and Apes. Kegan Paul, London.

THE SOCIAL LIFE OF MONKEYS, APES AND PRIMITIVE MAN [1]

BY MARSHALL D. SAHLINS
University of Michigan

INTRODUCTION

THIS study compares societies of infrahuman primates with the most rudimentary of documented human social systems. The objectives are to describe general trends in primate social organization leading to human society, and to delineate the major advances of the latter, cultural society, over precultural society.

For comparative materials, we rely on field studies of monkey and ape social behavior, supplementary observations of these animals in captivity, and ethnographic accounts of simple hunters and gathers. The quality and quantity of published studies of free ranging subhuman primate societies do not provide comfortable support for weighty generalizations. Aside from the anecdotal literature, we have only Carpenter's accounts of spider monkeys, rhesus, howling monkeys, and gibbons; Nissen's chimpanzee material; Zuckerman's observation of baboons (perhaps biased by his captivity studies); the as yet incomplete reports on the Japanese monkey; and some peripheral notes on the African red-tailed monkey by Haddow, and on the gorilla by Schwab.[2] *Considering this, our interpretations of subhuman primate social behavior are entirely provisional.* The data on primitive food gathers are more abundant. We include in our comparison the following primitive societies: Australian Aborigines, Tasmanians, Semang, Andamanese, Philippine and Congo pygmies, Bushmen, Eskimo, Great Basin Shoshoni,

[1] I thank Professors M. H. Fried, F. P. Thieme, S. L. Washburn, E. R. Service, J. N. Spuhler and L. A. White for their useful criticisms of an earlier version of this paper. I am especially indebted to White's articles touching subjects dealt with here (White, 1949, Chapters II, III, VI, XI). Needles to say, the above named do not necessarily agree with all statements contained herein.

[2] The following provide the field data on subhuman primates used in this paper: Carpenter, 1934; 1935; 1940; 1942c; Collias and Southwick, 1952; Haddow, 1952; Hooton, 1942 (report of Schwab's observations); Imanishi, 1957; Nissen, 1931; and Zuckerman, 1932. Unless quoting or making special points, these works will not be specifically cited hereafter.

Naskapi, Ona and Yahgan.[3] It is asumed that these societies parallel early cultural society in general features. This is simply an assumption of order and regularity. The technologies and low productivity of modern hunters and gathers resemble the archaeologically revealed productive systems of early cultures. Granting that a cultural social system is functionally related to its productive system, it follows that early human society resembles rudimentary, modern human society. This reasoning is supported by the large degree of social similarity among the present hunters and gathers themselves, despite the fact that some of them are as historically distant from each other, as separated in contact and connection, as the paleolithic is separated from modern times. Further, simply because many food gatherers have been driven into marginal areas, they are not thereby disqualified from consideration. There still remain strong social resemblances between marginal peoples, such as Bushmen, Ona, and Eskimo, and those found in isolated, but otherwise not ecologically marginal areas, such as many Australian groups and the Andaman Islanders.

A comparison of subhuman primate and primitive society must recognize the qualitative difference between the two (see White, 1949: Chapter III). Human society is cultural society; the organization of organisms is governed by culture traits. The social life of subhuman primates is governed by anatomy and physiology. Variations in human society are independent of, and are not expressions of, biological variations of the organism. Variations in primate society are direct expressions and concomitants of biological variation. Nissen writes: ". . . with one notable exception the phylogenetic course of behavioral development has been gradual . . . it has been a continuous affair, proceeding by quantitative rather than qualitative changes. The one exception is that which marks the transition from the highest nonhuman primates to man. . . . At this point a new ' dimension ' or mode of development emerges: culture." (Nissen, 1951: 426).

It follows that assertions of specific phylogenetic continuities from

[3] The following provide the field data on hunters and gathers used in this paper: Birket-Smith, 1936; Bleek, 1928; Boas, 1888; Cooper, 1946a; 1946b; Elkin, 1954; Forde, 1934; Gusinde, 1955; Leacock, 1954; 1955; Lothrop, 1928; Man, 1885; Putnam, 1953; Radcliffe-Brown, 1930-1931; 1948; Rink, 1875; Roth, 1890; Schapera, 1930; Schabesta, 1933; n. d.; Sharp, 1934-1935; Spencer and Gillen, 1927; Steward, 1938; Vanoverbergh, 1925; Warner, 1937; and Weyer, 1932. Unless quoting or making special points, these works will not be specifically cited hereafter.

anthropoid to primitive society must be summarily rejected, such as Yerkes' suggestion that delousing among primitives is a genetic survival of primate grooming, an activity which, Yerkes writes, also led to: "tonsorial artistry, nursing, surgery, and other social services of man." (Yerkes, 1933: 12). In the same vein is Kempf's identification of the presenting behavior of a subordinate rhesus monkey with human prostitution (Kempf, 1917, cf. Miller, 1931). Furthermore, the terminology of cultural social and political organization should be disavowed when describing infrahuman primate society. The cultural anthropologist justifiably shudders when he reads of "clans," "communism," and "socialism," among howler monkeys, or, "despots," "tyrants," "absolutism," and "slavery," among baboons.

The determinants of sociability are different in cultural and precultural society. We find useful Zuckerman's contention that social organization in general is based upon, "three main lines of behavior—search for food, search for mates, avoidance of enemies." (Zuckerman, 1932: 17). Of these factors, the sexual one appears to be primary in the genesis of subhuman primate society: "The main factor that determines social groupings in subhuman primates is sexual attraction." (*Ibid*: 31). "The emergence of this feature [*i. e.*, of continuous sex activity] into prominence in their behavior has created primate society." (Chance and Mead, 1953: 415). It was the development of the physiological capacity to mate during much of, if not throughout, the menstrual cycle, and at all seasons, that impelled the formation of year round heterosexual groups among monkeys and apes.[4] Within the primate order, a new level of social integration emerges, one that surpasses that of other mammals whose mating periods, and hence heterosexual groupings, are very limited in duration and by season. Certainly, defense against predation is also a determinant of subhuman primate sociability (Chance, 1955), but it, and the search for food, appear to be secondary to sex. The influence of sexual attraction in promoting solidarity among subhuman primates has been noted in the field. Carpenter has observed of the howlers: "With repetition of the reproductive cycle in the female and with uninterrupted breeding throughout the year, the process of group integration through sexual behavior is repeatedly operative, establishing and reinforcing intersexual social bonds." (Carpenter, 1934: 95; for a common alternate view, see Yerkes and Yerkes, 1935: 979.)

[4] The Japanese monkey, *Macaca fuscata*, northernmost of all subhuman primates, is reported to have a breeding season (Imanishi, 1957).

Sexual attraction remains a determinant of human sociability. But it has become subordinated to the search for food, to economics. A most significant advance of early cultural society was the strict repression and canalization of sex, through the incest tabu, in favor of the expansion of kinship, and thus mutual aid relations (Malinowski, 1931; Tylor, 1888; White, 1949: Chapter XI). Primate sexuality is utilized in human society to *reinforce* bonds of economic and to a lesser extent, defensive alliance. "All marriage schemes are largely devices to check and regulate promiscuous behavior in the interest of human economic schemes." (Miller, 1931: 402). This is not to underestimate the importance of primate sexuality in determining certain *general* characteristics of human society. If culture had not developed in the primate line, but instead among creatures practicing external fertilization, marriage and rules of exogamy and endogamy would not be means of establishing cohesive groups in cultural society. But, in the transition from subhuman to human society, cooperation in subsistence activities became the dominant cause of solidarity, avoidance of enemies a secondary cause, while sex became simply a facilitating mechanism.

These propositions are best documented by detailed consideration of primate and primitive society, wherein the differences in causes of sociability will be seen to pervade the comparison, and to turn generic continuity into specific discontinuity.

SUBHUMAN PRIMATE AND HUMAN PRIMITIVE SOCIETIES

Territoriality is one of a number of common features of subhuman primate and primitive social behavior. It is also general among lower vertebrates—perhaps it is a universal characteristic of society. Territoriality arises from competition over living conditions, and has the selective advantage of distributing the species in its habitat so as to maintain population density at or below its optimum (Allee, *et al.*, 1949: 411 f; Bartholomew and Birdsell, 1953: 485).

Territorial relations among groups of subhuman primates of the same species are generally exclusive. Except for the few animals that are driven out of one group and may attach themselves to another, primate societies are usually semi-closed societies. (Carpenter, 1942a: 187). Each horde has a focus of favorite feeding and resting places to which it is often deflected by contact with other groups. Contact between groups of the same species at territorial borders is generally competitive and antagonistic, sometimes violently so. Subhuman primate groups apparently have little tendency to federation. They, "*do not have supergroup social mechanisms. . . .* Kinship relations are not operative and inbreeding is the rule rather than the exception. . . ." (Carpenter 1954: 98). Carpenter finds the origin of intergroup cooperation characteristic of primitive tribalism, "difficult to trace in subhuman primates which I have studied." (1940: 163).

Primate territorial relations are altered by the development of culture in the human species. Territoriality among hunters and gatherers is never exclusive, and group membership is apt to shift and change according to the variability of food resources in space and time. Savage society is open, and corresponding to ecological variations, there are degrees of openness: 1. Where food resources are evenly distributed and tend to be constant year to year, territory is clearly demarcated and stable, and, considering the nucleus of males, so is group membership. The Ona and most Australian groups are representative of this condition. Stability of membership is effected through customary rules of patrilocal residence, which is usually coupled with local exogamy. Patrilocal residence confers the advantages of continuity of occupation for hunters in areas with which they are familiar. 2. Where food resources are evenly distributed during some seasons, and variably abundant during others, exclusiveness of territory and membership are only seasonal. Local groups of Central Kalahari Bushmen, for example, remain fixed in territories focused around water holes during the dry season; whereas, in the rainy reason, such groups mingle and hunt together. Similarly the Semang have seasonal territories fixed by the distribution of the durian tree; after the harvest, territories and exclusive groups dissolve, to be reconstituted at the next durian season. Under these conditions, patrilocal residence and local exogamy are preferred, but not strict, rules, and local endogamy and matrilocality occur. The Andaman Islanders with relatively fixed territories, a tendency toward local exogamy, and no residence rule, and the Yahgan, who tend to be locally exogamous, patrilocal, and territorially exclusive, may also fall into this ecological type. However, data available for classification of these groups are inadequate. 3. Finally, there are food gatherers among whom rules of territoriality are *de facto* nonexistant, and local group composition highly variable. These occupy areas where food resources vary in local abundance seasonally and annually. Families coalesce and separate *ad hoc* corresponding to accessibility of supplies. Postmarital residence may be in the group of either spouse, and band aggregates are agamous—there are no rules.

The Great Basin Shoshone, the Eskimo, and the pre-fur trade Naskapi fall into this category.[5]

Territoriality among hunters and gatherers is sometimes maintained by conflict. There appears to be a direct relationship between intensity of intergroup feud and exclusiveness of territory and membership. Thus trespass is strongly resented and interband feuds are relatively frequent among Australians and the Ona; whereas, where territoriality is weak, as among Eskimos and Shoshoni, the concept of trespass is naturally poorly developed, and fighting consists of squabbles between particular families. However, in all cases the outcome of trespass depends on the previous relations between neighbors, and these are usually *friendly*. Even among the Australians, adjacent groups would be allowed to hunt in a band's territory if in need; there is "no constant state of enmity" between neighbors (Spencer and Gillen, 1927, i: 58). Everywhere, no matter how strict the rules of territory, constant visiting between bands prevents the development of closed groups. And everywhere exclusiveness is easily broken down if there is some food windfall, or if food is differentially abundant in adjacent locales.

Hunters and gatherers live in relatively open groups between which relations are usually friendly; infrahuman primates of the same species live in relatively closed groups between which relations are usually competitive. The invention of kinship and the incest tabu of cultural society are responsible for this difference. Through marrying out, friendly, cooperative relations are established between families. When exogamy can be extended to the local group, cooperation between bands is effected (Tylor, 1888; White, 1949; Chapter XI). Intermarrying Australian bands are described as, "*a family* of countries bound together by those sentiments which function between members of a family and its near relations." (Elkin, 1954: 81; emphasis ours). It is the kinship

[5] Philippine and Congo pygmies are presently of the first type: rigid territory, patrilocal, exogamous. However, both live in symbiotic, subservient relation to agricultural peoples; and, at least for the Congo Twides, territorial exclusiveness is clearly a function of boundaries between patron Bantu villages (Putnam, 1953; Schabesta, 1933; Gusinde would disagree, 1955: 23). The Heikum Bushmen, also in symbiotic, subservient status to patrilineal Hottentot and Bantu groups, are another instance of the same thing. Unlike other Bushmen, the Heikum live in well defined territories and practice strict patrilocality and local exogamy (Schapera, 1930: 34-35, 38, 83, 94f). Leacock's (1955) Naskapi studies suggest that bilocal, nonterritorial, unstable bands commonly become so formalized under outside influences of this general sort.

ethic of mutual aid that permits populations of hunters and gatherers to shift about according to the distribution of resources. Kinship is thus selectively advantageous in a zoological sense; it permits primitives to adjust to more variable habitats than subhuman primates (see Carpenter, 1955: 93). Moreover, the kinship relations between groups, and the ceremonies and exchanges of goods that frequently accompany interband meetings, give rise to a further social development: tribalism. Common custom, common dialect, a name and a feeling of unity are created among otherwise independent groups. The stage for further political evolution is thereby set.

We turn now to the internal organization of subhuman primate and primitive human societies.

The subhuman primate horde varies in size from an average of four animals among the gibbon to several hundred baboons (Carpenter, 1942a). Group size is not correlated with suborder differences, except that great ape hordes are generally at the lower end of the primate range; orangutan groups are apparently as small as gibbons' and chimpanzees average 8.5 per group. The horde may remain together at all times, or may disperse, during daytime feeding, into segments of various constitution—mating groups, female packs, male packs—concentrating at night resting places. Howling monkeys typically travel together; spider monkey and baboon groups are instances of segmented hordes.

With the exception of the gibbon, the primate horde characteristically contains more adult females than adult males. In observed wild groups the ratio ranges from nearly 3:1 for howlers, to 1:1 for gibbons. The ratio for spider monkeys, 1.6 females per male, is probably near the average for the primates. The usual inequality apparently reflects the degree of dominance and competition among males for female sexual partners.[6] There are almost always unmated males, either periphereally attached to heterosexual groups, or existing outside the horde.

We have argued that sexual attraction is the primary cause of subhuman primate sociability. Indeed, in many cases the entire horde is a single reproduction unit or mate group. But there are significant species variations in the constitution of hordes and mate groups. There appears to be a progressive development within the primate order from promiscuous relations within the group to the establishment of exclusive sex partnerships, one of which comprises the nucleus of a horde. Since the

[6] Washburn has suggested (personal communication) that the unequal ratio may in good part be due to the faster maturation rate for females.

young remain attached to adult females throughout, the highest forms of primate mate groups resemble the elementary human family in composition (Carpenter, 1942a: 186; Chance and Mead, 1953: 418; Yerkes and Yerkes, 1929: 566). The emergence of exclusive, independent mate groups takes the following steps: 1. Among New World howler and spider monkeys observed in free-ranging conditions, the only stable relations within a horde are between females and their young. Only when females are in the oestrus period of the menstrual cycle do they leave the female-offspring pack and become attached to specific males, and then not exclusively, but to several in succession. The mated pair is a temporary, non-exclusive unit. 2. Old World monkeys develop more permanent sex partnerships. Rhesus shows the trend toward exclusiveness. Again in rhesus, the female-young pack is a separate unit, and sex partnerships are only temporarily established while a female is in heat. However, for every female, the succession order among her male partners corresponds to their dominance position. Therefore, dominant males have all females in oestrus, while subordinate males are excluded from some when there are not enough to go around. The mate group of the Japanese monkey, in the same genus as rhesus, is very similar. 3. The baboon mate group is a simple extrapolation from rhesus. The steps involved are: the exclusion of subordinate males from sexual relations with females, and the development of constant association between a dominant male, females and young. The baboon mate group is a permanent, exclusive relationship between a dominant male and his several females, the young following their mothers. Subordinate males may remain attached to the group on its fringes, or form unisexual bands, The baboon horde consists of several such mate groups and male bands, each relatively independent. 4. The ape horde tends to be composed of a single, independent mate group of the baboon type. In the gibbon, this consists of one male, one female plus young. The evidence from the other anthropoids is not conclusive; however, gorilla, orangutan and chimpanzee hordes apparently consist of a single mate group of one male, two or more females and their young, and perhaps subordinate males.

The emergence of exclusive, permanent mate groups among higher primates is explained by the progressive emancipation of sexual behavior from hormonal control running through the order (Beach, 1947). In monkeys, copulation outside the female's fertile period is relatively rare. In apes—more commonly so, in man—sex is freed from hormonal regulation, being subject instead to cortical and social control.[7] The oft-made alternative assertion that the development of the family is due to increasing duration of infant dependency is not supportable. With one minor exception, subhuman primate males are never significantly involved in rearing the young, save in retrieving the fallen, and indirectly as group defenders. The exceptional case of the male Japanese monkey that rears the older infant if a female has two in succession is not significant here, since there is no family-type group involved. This behavior was observed only in one of a number of hordes of Japanese monkeys, and is entirely unique among subhuman primates. Baboon and rhesus males are known to have killed young in the course of sexual attacks upon their mothers. A long dependency period cements mother-offspring relations but not necessarily father-mother-offspring relations. Only when there is an economic division of labor by sex can infant dependency produce this effect (Dole, n. d.). In subhuman primates there is no sexual division of economic labor.

Social relations within the subhuman primate horde vary according to the age, sex and dominance statuses of the interacting animals. Leaving aside dominance for a moment, most social interaction can be adequately described by utilizing three elemental status categories: adult male, adult female, and young (Carpenter, 1942a: 180). Interaction of animals of these categories produce 6 "qualitatively distinct" social relations: male-male; female-female; male-female; male-young; female-young; young-young (Carpenter, 1940: 126). The content of most of these relations can be inferred from the preceeding and succeeding discussion. It will be seen that age and sex difference remain important social distinctions among primitive hunters and gatherers.

Dominance statuses are found among all known monkeys and apes, as well as many lower vertebrates (Carpenter, 1942a; Maslow, 1936a). Dominance is established by competition for mates, for food, for position in progression, and the like—". . . in every known typical grouping of monkeys and apes there is persisting competition for priority rights to incentives." (Carpenter, 1942a: 192). Conflict is often particularly prominent among males over females in oestrus. Males are usually dominant over females. [This last is subject to exception when females are in heat (Crawford, 1940; Nowlis, 1942; Yerkes, 1940). Dominance

[7] This is a crucial physiological change for a number of reasons, not the least of which is that it makes intelligible how, when culture developed, primate sexuality was so radically subordinated to other ends.

has been experimentally raised in primates and other vertebrates by injections of male sex hormones (Clark and Birch, 1945; Birch and Clark, 1946).] Dominance status affects behavior in every aspect of social life: play, feeding, sex, grooming, competition between groups, and it even determines the spatial relations of animals within the horde (Carpenter, 1942a; Chance and Mead, 1953; Maslow, 1936a; 1936b; Maslow and Flanzbaum, 1936; Nissen, 1951). Dominance among paired animals is easily determined experimentally by introducing a series of food pellets to which there is limited access, and noting which animal consistently appropriates them (Maslow, 1936a; Nissen, 1951: 447; Nowlis, 1941a; 1941b). Nowlis' food appropriation experiments, performed with differentially satiated animals, show that dominance is a social behavior arising from conflict—not a simple, independent drive for "prestige," as is sometimes claimed (Nowlis, 1941b; 1942).

Maslow contends that the quality of dominance varies among New and Old World monkeys and the apes, and that variation in dominance quality is correlated with differences in social organization (Maslow, 1940). Platyrrhines, according to Maslow, show the greatest indifference in social relations. Dominance is "tenuous," "non-contactual," frequently not expressed, and often ascertained in the laboratory only with difficulty. In contrast, "Catarrhine dominance is rough, brutal and aggressive; it is of the nature of a powerful, persistent, selfish urge that expresses itself in ferocious bullying, fighting and sexual aggression." (Maslow, 1940: 316). Weak and sick animals are attacked; in competition over food, a subordinate animal would starve. Chimpanzees however, show "friendly dominance." Dominant animals protect subordinates, never attack them except in the form of rough play. Crawford (1942) in one laboratory experiment noted that only 0.4% of chimpanzee social behavior could be described as aggressive, and other field and captivity data generally bear out Maslow's hypothesis of suborder differences in dominance quality (cf. Carpenter, 1934; 1937; 1942a; Collias and Southwick, 1952; Gillman, 1939: Harlow and Yudin, 1933; Maslow, 1936b; Warden and Galt, 1943; Yerkes and Yerkes, 1929; 1935; Zuckerman, 1932; for an exception, see Haddow, 1952).

There appear to be correlated differences in the grooming behavior of the suborders. Yerkes advances the notion that social, as opposed to self grooming, increases in frequency from prosimian through authropoid ape (Yerkes, 1933). Grooming serves a biological function in removing parasites and the like, but this is nearly equally accomplished by self or social grooming. Therefore, social grooming takes on added significance as a "social service" in the higher primates. Moreover, grooming in higher primates is not only social, but reciprocal. Social grooming among wild gibbons has been noted to be frequent and reciprocal. By contrast, evidence from field and laboratory indicates that social grooming is comparatively infrequent among New World monkeys. It is difficult to say from present evidence that social and reciprocal grooming increase from Old World monkeys to anthropoids, but on the whole, Yerkes assertions are supportable (cf. Carpenter, 1935; Crawford, 1942; Maslow and Flanzbaum, 1936; Warden and Galt, 1943).

The emergence of reciprocal social behavior and the progressive tempering of dominance relations are significant trends in the primate line—trends which, we shall note, are continued in primitive society. On the other hand, economic teamwork and mutual aid are nearly zero among subhuman primates, including anthropoids. Spontaneous cooperation—as opposed to one animal helping another—has not been observed among them. Chimpanzees have been trained to solve problems cooperatively, but fail to do so without tuition (Crawford, 1937). Monkeys apparently cannot even be taught to cooperate (Warden and Galt, 1943). Spontaneous teamwork presupposes symboling: "Teamwork makes intellectual demands of the same order as those made by language. Psychologically, it may, in fact, be difficult to distinguish between the two." (Hebb and Thompson, 1954: 540). Nissen and Crawford's elaborate experiment showed that chimpanzees share food pellets and tokens, although sharing was much less frequent than not, and was evidently non-reciprocal (Nissen and Crawford, 1936). However, in a similar experiment using animals tested for dominance, Nowlis observed that every case of sharing (in 480 trials there were 80 instances of sharing and, in these, one tenth of the food available was shared) was from subordinate to dominant; dominants never gave food to subordinates (Nowlis, 1941). Therefore, "food sharing" here is a function of previous dominance competition and actually indicates monopolization, not pooling, of a limited supply (Nissen, 1951:447).

We now consider the social system of bands of hunters and gatherers. The usual band contains 20 to 50 people (Steward, 1955: 146), but during poor seasons it may fragment into small family groups. We have already looked at band structure. Corresponding to ecological conditions, bands range in composition from an enlarged patrilocal family to a congeries of variably related nuclear families. Almost all members will be kinsmen, and the etiquette of kinship behavior regulates social life.

In Australia, for example, kinship: ". . . regulates more or less definitely the behavior of an individual to every person with whom he has any social dealings whatsoever." (Radcliffe-Brown, 1930-1931: 43).

The family is the only social unit inside the band; where bands are unstable, it is the major form of social organization. Among primitives, the division of economic labor by sex is fundamental to the family and makes marriage an economic alliance—or, in Westermarck's terms, ". . . marriage is something more than a regulated sexual relation. It is an economic institution." (Westermarck, 1922, i: 26). The complementary economic roles of the sexes determine certain qualities of primitive marriage. First, it is a necessity for all adults; the unmated adult male of the subhuman primate horde rarely has a counterpart in primitive bands. Secondly, polygamy is usually economically impractical; monogamy is prevalent. Finally, as indicated, stable heterosexual relations are not simply determined by sexual attraction, but by economics. Sex is easily had in many hunting and gathering groups, both before and beside marriage, but such neither necessarily establishes the family nor destroys it. Legal rights to sexual privileges of spouses may be waived in favor of socio-economic advantages, as in wife lending. The very rules of exogamy and incest prevent the formation of the family on a basis of simple sexual attraction. Steward's statement of the economic basis of the Shoshoni marriage can be duplicated from accounts of other simple societies: "Marriage was an economic alliance in a very real sense . . . a union which brought into cooperation the complementary economic activities of the sexes—a person could not, in the interest of self-preservation, afford to remain long single . . . the role of exclusive sex privileges in matrimony seems to have been secondary." (Steward, 1938: 242). Compare Radcliffe-Brown on the Australians: "The family is based on the co-operation of man and wife, the former providing the flesh food and the latter the vegetable food . . . this economic aspect of the family is a most important one. . . . I believe that in the minds of the natives themselves this aspect of marriage, *i. e.*, its relation to subsistence, is of greatly more importance than the fact that man and wife are sexual partners." (Radcliffe-Brown, 1930-1931: 435).

In man, therefore, primate sexuality has been brought under cultural control; it has become, in part, a means to other ends. Another aspect of primitive marriage teaches the same lesson. Unlike primate unions, created and maintained in conflict, primitive marriage is a powerful factor in interfamilial and interband alliance. Again, as Steward writes of the Shoshoni, and one can find countless ethnographic paraphrases of this truism: "Marriage was more a contract between families than between individuals." (Steward, 1955: 118). Among hunters and gatherers, marriages are frequently arranged (or at least approved) by the families rather than the spouses. Considerations often pass between the groups, thus setting the pattern for future cooperation. Women may be exchanged between groups, or intermarriage between certain parties preferred and repeated, solidifying both the marriages and group relations. The alliance between families may be paramount to the extent that a marriage can survive the death of one of the partners, he or she being replaced through the levirate or sororate.

There is an outstanding implication of these characteristics of primitive marriage and the family. Given the division of labor by sex and the formation of domestic units through marriage, it follows that sharing food and other items, rather than being non-existent, as among monkeys and apes, is a *sine qua non* of the human condition. Food sharing is an outstanding functional criterion of man. In the domestic economy of the family there is constant reciprocity and pooling of resources. And, at the same time that kinship is extended throughout the band of families, so are the principles of the domestic economy. Among all hunters and gatherers there is a constant give and take of vital goods through hospitality and gift exchange. Everywhere, generosity is a great social virtue. Also general is the custom of pooling large game among the entire band, either as a matter of course, or in times of scarcity. Where kinship is extended beyond the local group by inter-band marriage, so are reciprocity and mutual aid. Goods may pass over great distances by a series of kinship transactions. Trade is thus established. Hunters and gatherers are able to take mutual advantage of the exploitation of distant environments, a phenomenon without parallel in the primate order. This emphasis on generosity, on mutual aid, and the attribution of social prestige for generosity, stand in direct opposition to the tendency among primates to monopolize vital goods. Perhaps the elaborated emphasis on sharing among primitives is to be partially understood as a cultural means for overcoming primate tendencies in the opposite direction.

In the system of social status, there is a generic continuity between primate and primitive society. Qualitative social differences of sex and age that are marked in subhuman primate groups are major principles of status and role allocation among hunters and gatherers. The division of labor by sex is an example. So is the pervasive recognition of sex, seniority and generation in kinship behavior and terminology (*e.g.*, see Radcliffe-Brown, 1948).

There is also some continuity in dominance status. Leadership falls to men among hunters and gatherers, although, what is possibly different from subhuman primates, it is especially the elders that are respected. There are, however, important qualitative differences in dominance relations among hunters and gatherers and subhuman primates. Each primitive band usually has an elderly headman. The respect accorded him and other elders is not due to their physical ability to appropriate a limited supply of desired objects. (They may be preferentially treated in communal food distribution, but this is another thing.) Quite the opposite from subhuman primates, a man must be generous to be respected. Prestige among hunters and gatherers can be estimated by noting who gives away the most—precisely the reverse of the test for dominance status among subhuman primates. The position of the head man rests primarily on superior knowledge of game movements, water and other resources, ritual and other things which govern social life. Thus Boas pointed out that there is a direct relation between the authority of Eskimo headmen of various groups and the distance and difficulties involved in traveling between winter and summer hunting grounds (Boas, 1888). But such knowledge alone cannot breed power. The leader of the band has no means to compel obedience. He is commonly described as ruling through "moral influence." A Congo pymgy leader bluntly told Schabesta, "There would be no point in his giving orders, as nobody would heed them." (Schabesta, 1933: 104). Steward comments that the title, "talker," given to a Shoshoni leader, "truly designates his most important function." (Steward, 1938: 247). The leader of the Central Eskimo camp is picturesquely referred to as *isumataq*, 'he who thinks' (for the others) (Birket-Smith, 1936: 148). In all bands of hunters and gatherers, the heads of the separate families exercise more control than the informal headmen over the whole. Compared to infrahuman primates, ranking hierarchies and dominance approach zero among hunters and gatherers. Yet, where all interact as kinsmen, and sharing a scanty food supply replaces conflict over it, this is expectable.

SUMMARY AND CONCLUSIONS

The transition from subhuman primate society to rudimentary cultural society was at the same time a process of generic continuity and of specific discontinuity. If culture had not developed among a species of primates, but among animals of different behavioral characteristics, then the forms and development of cultures would be basically different. The social behavior of primates is the foundation of some general features of human society. On the other hand, no specific trait of cultural society, even in its most rudimentary state is, in both form and functioning, a direct survival of some specific trait of primate social behavior. This discontinuity is due to the fact that subhuman primate society is a direct expression of the physiology of the species operating in a given environment; whereas cultural traits govern the social adaptation of the human primate. The development of culture did not simply give expression to man's primate nature, it replaced that nature as the direct determinant of social behavior, and in so doing, channeled it—at times repressed it completely. The most significant transformation effected by cultural society was the subordination of the search for mates—the primary determinant of subhuman primate sociability—to the search for food. In the process also, economic cooperation replaced competition, and kinship replaced conflict as the principal mechanism of organization.

What are the generic continuities? Territoriality is one. But similarities such as these are common to a wide variety of societies, including those of lower vertebrates. A more restricted continuity is the utilization of the powerful social functions of primate sexuality in human social organizations. To repeat an earlier observation, it is only against the unique background of primate sexual behavior that one fully understands why marriage and marriage rules are general mechanisms for integrating cooperative human groups. Another generic survival of simian society is the allocation of social functions on the lines of sex and age among hunters and gatherers.

Of particular interest are social advances within the primate order, upon which cultural society directly elaborated. Here the cultural developments appear capstones to trends which had begun to unfold in precultural conditions. In the primate line the exclusive mate group appears to have developed out of the promiscuous horde. In the transformation to the human family, the anthropoid mate group was altered more in function than in form. A second primate advance is the development of reciprocal social services in grooming. Generalized to food sharing, reciprocity is basic to cultural society. Thirdly, there is the softening of dominance relations among higher primates. In primitive society, dominance or prestige is especially associated with service to the group. (*See* also Hallowell, 1956, on generic similarities).

The most significant advances in the early evolution of cultural society can be deduced by comparison of primate and primitive groups. To us these advances appear to be: 1. the division of labor by sex and the establishment of the family on this basis; 2. the invention of kinship; 3. the incest prohibition and its extension through exogamy, thus extending kinship; 4. the overcoming of primitive competition over food in favor of sharing and cooperation; and 5. the abolition of other primate conflicts leading to the establishment of dominance hierarchies.

These 5 are complementary; nothing is said here of their order of appearance or relative significance. It is claimed that they are great triumphs of early culture. These developments are necessary social counterparts of the continuous tool activity which enabled man to become the dominant form of life.

LITERATURE CITED

ALLEE, W. C., ALFRED E. EMERSON, ORLANDO PARK, THOMAS PARK AND KARL P. SCHMIDT 1949 Principles of Animal Ecology. W. B. Saunders Company, Philadelphia and London.

BARTHOLOMEW, GEORGE A. AND JOSEPH B. BIRDSELL 1953 Ecology and the protohominids. Amer. Anthrop. *55*: 481-498.

BEACH, FRANK A. 1947 Evolutionary changes in the physiological control of mating behavior in mammals. Psychological Review *54*: 297-315.

BIRCH, HERBERT S. AND GEORGE CLARK 1946 Hormonal modification of social behavior (II). The effects of sex hormone administration in the social dominance status of the female-castrate chimpanzee. Psychosomatic Med. *8*: 320-331.

BIRKET-SMITH, KAJ 1936 The Eskimos. Methuen and Company, London.

BLEEK, D. F. 1928 The Naron, A Bushman Tribe of the Central Kalahari. Cambridge University Press, Cambridge.

BOAS, FRANZ 1888 The Central Eskimo. Smithsonian Institution, Bureau of Ethnology, Annual Report *6*: 401-669.

CARPENTER, C. R. 1934 A field study of the behavior and social relations of howling monkeys. Comparative Psych. Monogr. *10* (2): 1-168.

——— 1935 Behavior of red spider monkeys in Panama. J. Mammal. *16*: 171-180.

——— 1937 An observational study of two captive mountain gorillas. Human Biol. *9*: 175-196.

——— 1940 A field study in Siam of the behavior and social relations of the gibbon (*Hylobates lar*). Comparative Psychol. Monogr. *16* (5): 1-212.

——— 1942a Societies of monkeys and apes. *In* Biological Symposia 8: 177-204.

——— 1942b Characteristics of social behavior in non-human primates. Transactions N. Y. Acad. Sci. *4*: 248-258.

——— 1942c Sexual behavior of free ranging rhesus monkeys (*Macaca mulatta*) I and II. J. Comparative Psychol. *33*: 113-142; 143-162.

——— 1954 Tentative generalization on the grouping behavior of non-human primates. Human Biol. *26*: 269-276. Reprinted in James A. Gavan, ed., 1955, The Non-Human Primates and Human Evolution, Wayne University Press, Detroit.

CHANCE, M. R. A. AND A. P. MEAD 1953 Social behavior and primate evolution. Symposia of the Soc. for Experimental Biol. 7: 395-439. Academic Press, New York.

CHANCE, M. R. A. 1955 The sociability of monkeys. Man *55*(176): 162-165.

CLARK, GEORGE AND HERBERT BIRCH 1945 Hormonal modifications of social behavior 1. The effect of sex-hormone administration on the social status of a male-castrate chimpanzee. Psychocomatic Med. 7: 321-329.

COLLIAS, NICHOLAS AND CHARLES A. SOUTHWICK 1952 A field study of population density and social organization in howling monkeys. Proc. Amer. Philosophical Soc. *96*: 143-156.

COOPER, JOHN M. 1946a The Ona. *In* Handbook of South American Indians, Julian H. Steward, ed. Smithsonian Institution Bureau of American Ethnology Bul. *143*, vol. 1: 107-125.

——— 1946b The Yahgan. *In* Handbook of South American Indians, Julian H. Steward, ed. Smithsonian Institution Bureau of American Ethnology Bul. *143*, vol. 1: 81-106.

CRAWFORD, MEREDITH P. 1937 The cooperative solving of problems by young chimpanzees. Comparative Psych. Monogr. *14*(2): 1-88.

——— 1940 The relation between social dominance and the menstrual cycle in female chimpanzees. J. Comparative Psych. *30*: 482-513.

——— 1942 Dominance and social behavior for chimpanzees in a non-competitive situation. J. Comparative Psych. *33*: 267-277.

DOLE, GERTRUDE E. N. d. Primal Human Family Structure. ms.

ELKIN, A. P. 1954 The Australian Aborigines: How to Understand Them. 3rd ed., Angus and Robertson, Sydney.

FORDE, C. DARYLL 1934 Habitat, Economy and Society: A Geographical Introduction to Ethnology. Methuen and Company, London.

GILLMAN, JOSEPH 1939 Some facts concerning the social life of chacma baboons in captivity. J. Mammal. *20*: 178-181.

GUSINDE, MARTIN 1955 Pygmies and Pygmoids: Twides of Tropical Africa. Anthropological Quarterly *28*: 3-61.

HADDOW, A. J. 1952 Field and laboratory studies in an African monkey, *Cercopithecus ascanius schmidti*, Matschie. Proc. Zool. Soc. London, *122*: 297-394.

HARLOW, H. F. AND H. C. YUDIN 1933 Social behavior of primates: 1. Social faciliatation of feeding in the monkey and its relation to attitudes of ascendance and submission. J. Comparative Psych. *16*: 171-185.

HALLOWELL, A. IRVING 1956 The structural and functional dimensions of a human existence. Quarterly Review Biol. *31*: 88-101.

HEBB, D. O. AND W. R. THOMPSON 1954 The social significance of animal studies. *In* Handbook of Social Psychology, Gardner Lindsey, ed., Addison Wesley, Cambridge, Massachusetts, pp. 532-561.

HOOTON, EARNEST 1942 Man's Poor Relations. Doubleday, Doran, New York.

IMANISHI, KINJI 1957 Social behavior in Japanese monkeys, *Macaca fuscata.* Psychologia *1*: 47-54.

KEMPF, EDWARD J. 1917 The social and sexual behavior of infra-human primates with some comparable facts of human behavior. Psychoanalytic Review *4*: 127-154.

LEACOCK, ELEANOR 1954 The Montagnais 'Hunting Territory' and the Fur Trade. Memoir of the American Anthropological Association 78.

——— 1955 Matrilocality in a simple hunting economy (Montagnais-Naskapi). Southwestern J. Anthrop. *11*: 31-47.

LOTHROP, SAMUEL KIRKLAND 1928 The Indians of Tierra Del Fuego. Museum of the American Indian. Heye Foundation, New York.

MALINOWSKI, BRONISLAW 1931 Culture. Encyclopedia of the Social Sciences.

MAN, EDWARD HORACE 1885 On the aboriginal inhabitants of the Andaman Islands. *Reprinted from* J. Royal Anthropological Institute (1885). London, The Royal Anthropological Institute.

MASLOW, A. H. 1936a The role of dominance in the social and sexual behavior of infra-human primates: I. Observations at Vilas Park Zoo. J. Genetic Psych. *48*: 261-277.

——— 1936b The role of dominance in the social and sexual behavior of infra-human primates: III. A theory of sexual behavior of infra-human primates. J. Genetic Psych. *48*: 310-338.

——— 1940 Dominance-quality and social behavior in infra-human primates. J. Social Psych. *11*: 313-324.

MASLOW, A. H. AND SYDNEY FLANZBAUM 1936 The role of dominance in the social and sexual behavior of infra-human primates: II. An experimental determination of the behavior syndrome of dominance. J. Genetic Psych. *48*: 278-309.

MCCARTHY, F. D. 1938-1939 and 1939-1940 'Trade' in aboriginal Australia, and 'trade' relationships with Torres Straits, New Guinea and Malaya. Oceania *9*: 404-438; *10*: 80-104; 171-195.

MILLER, GERRIT S. 1931 The primate basis of human sexual behavior. Quarterly Review Biol. *6*: 379-410.

NISSEN, HENRY W. 1931 A field study of the chimpanzee: Observations of chimpanzee behavior and environment in Western French Guinea. Comparative Psych. Monogr. *8*(1): 1-122.

——— 1951 Social behavior in primates. *In* Comparative Psychology, C. P. Stone, ed. 3rd ed. Prentice-Hall, Englewood Cliffs, New Jersey, pp. 423-457.

NISSEN, H. W. AND M. P. CRAWFORD 1936 A preliminary study of food-sharing behavior in young chimpanzees. J. Comparative Psych. *22*: 383-419.

NOWLIS, VINCENT 1941a Companionship preference and dominance in the social interaction of young chimpanzees. Comparative Psych. Monogr. *17*(1): 1-57.

——— 1941b The relation of degree of hunger to competitive interaction in chimpanzee. J. Comparative Psych. *32*: 91-115.

——— 1942 Sexual status and degree of hunger in chimpanzee competitive interaction. J. Comparative Psych. *34*: 185-194.

PUTNAM, PATRICK 1953 The Pygmies of the Ituri Forest. *In* A Reader in General Anthropology, Carleton S. Coon, ed. Henry Holt, New York, pp. 322-342.

RADCLIFFE-BROWN, A. R. 1930-1931 The social organization of Australian tribes. Oceania *1*: 34-63; 206-256; 322-341; 426-456.

——— 1948 The Andaman Islanders. The Free Press, Glencoe, Illinois.

RINK, HENRY 1875 Tales and Traditions of the Eskimo. William Blackwood and Sons, Edinburgh and London.

ROTH, H. LING 1890 The Aborigines of Tasmania. Kegan Paul, Trench, Trubner & Company, London.

SCHAPERA, I. 1930 The Khoisan Peoples of South Africa. George Routledge and Sons, London.

SCHABESTA, PAUL 1933 Among Congo Pigmies. Hutchinson & Company, London.

——— N.d. Among the Forest Dwarfs of Malaya. Hutchinson & Company, London.

SHARP, LAURISTON 1934-1935 Ritual life and economics of the Yir-Yiront of Cape York Peninisula. Oceania *5*: 19-42.

SPENCER, SIR BALDWIN AND F. J. GILLEN 1927 The Arunta. 2 vols. Macmillan and Company, London.

STEWARD, JULIAN H. 1938 Basin-Plateau aboriginal socio-political groups. Smithsonian Institution Bureau of American Ethnology Bulletin 120.

——— 1955 Theory of Culture Change. University of Illinois Press, Urbana.

TYLOR, E. B. 1888 On a method of investigating the development of Institutions; applied to laws of marriage and descent. J. Royal Anthrop. Institute *18*: 245-269.

VANOVERBERGH, MORICE 1925 Negritos of Northern Luzon. Anthropos *20*: 147-199; 399-443.

WARDEN, C. J. AND WILLIAM GALT 1943 A study of cooperation, dominance grooming, and other social factors in monkeys. J. Genetic Psych. *63* 213-233.

Warner, W. Lloyd 1943 A Black Civilization. Harper and Brothers, New York and London.

Westermarck, Edward 1922 The History of Human Marriage. 3 vols. Allerton, New York.

Weyer, Edward Moffat 1932 The Eskimos. Yale University Press, New Haven.

White, Leslie A. 1949 The Science of Culture. Farrar, Straus and Company, New York.

Yerkes, Robert M. 1933 Genetic aspects of grooming, a socially important primate behavior pattern. J. Social Psych. *4*: 3-25.

——— 1940 The social behavior of chimpanzees: Dominance between mates in relation to sexual status. J. Comparative Psych. *30*: 147-186.

Yerkes, Robert M. and Ada Yerkes 1929 The Great Apes. Yale University Press, New Haven.

——— 1935 Social behavior in infrahuman primates. *In* A Handbook of Social Psychology, Carl Murchison ed., Clark University Press, Worchester, Massachusetts, pp. 973-1033.

Zuckerman, Solly 1932 The Social Life of Monkeys and Apes. Harcourt, Brace and Company, New York.

——— 1933 Functional Affinities of Man, Monkeys, and Apes. Harcourt, Brace and Company, New York.

SUMMARY REVIEW

BY LESLIE A. WHITE
The University of Michigan

MY ROLE, in writing a summary review of the preceding papers is not, as I conceive it, one of merely passing judgments upon them, saying that this one is good, that one not-good, and so on. Rather, I think of my function as making such comments and observations about these papers as would help the reader toward greater comprehension and finer appreciation of the contributions, individually and collectively. But judgment and evaluation on my part cannot be wholly eliminated since anything I say must proceed from a point of view. Where I differ with an author of any one of the essays it will be understood, I hope, that this is merely an expression of differences of premise, interpretation, and conclusion that characterize discussions of open questions everywhere and to which all workers in the field of science are entitled.

The essays express two quite different conceptions of culture. Spuhler and Sahlins restrict *culture* to the human species; there is a qualitative difference between the cultures of man and the behavior of subhuman species. Hockett thinks of culture as behavior that "is not only learned but *taught*"; he finds "culture of a rather thin sort among the hominoid apes, and even stronger evidence in the case of waterfowl." Harlow speaks of both human and subhuman cultures. Washburn has both culture and proto-culture. The former is, apparently, limited to *Homo sapiens*; the latter was possessed by earlier species or genera of man. Washburn does not say whether the difference between proto-culture and culture "as we know it" is merely quantitative or whether it is qualitative as well; he merely says that the two are "very different." In one instance at least, if not throughout his paper, Washburn seems definitely to equate culture with tool-using: "Our hand is the result of at least half a million years of tool use. . . . [Its] uniqueness . . . is the result of culture." Gerard does not concern himself with the conception of culture as such.

I subscribe to the conception and definition of culture which limits it to the human species. If we include behavior of nonhuman species within the concept culture what term are we to use to designate behavior that is peculiarly human? It would seem to be undesirable to use "human culture" and "subhuman culture" if there is a fundamental (qualitative) difference between the two. To be sure, there are those who deny that man's mind differs fundamentally and qualitatively from that of subman. I believe, however, that the uniqueness of man's mind has been demonstrated, and I have tried to set forth and make explicit this demonstration and to define the distinction between human and nonhuman in my essay, "The Symbol: the Origin and Basis of Human Behavior" (in White, 1949). Man has an ability that no other animal possesses. For want of a better term we have called it the ability to symbol, namely, the ability freely and arbitrarily to impose meanings upon things and events and the ability to grasp and appreciate meanings thus determined. Symboling is traffic in meanings that cannot be comprehended with the senses. Articulate speech, the meaning of holy water, the significance of fetiches, etc., are examples of this kind of behavior. We would, therefore, define human behavior and culture in terms of the ability to symbol. We would distinguish human behavior from culture by defining the former as things and events dependent upon symboling considered, for purposes of scientific interpretation, in a somatic context; culture consists of things and events dependent upon symboling considered in an extrasomatic context (White, 1959).

We have, then, to consider, in a symposium on the evolution of man's capacity for culture, those biological developments that prepared the way for and made possible a way of life consisting of ideologies, institutions, customs, rituals, and technologies.

But when we say "evolution of the capacity for culture" do we mean the culture that has actually been realized by the human species—"culture as we know it," as Washburn puts it—or "other conceivable kinds of culture"? The fact that man is carnivorous-omnivorous is significant for the kind of culture that he has developed. But it is quite conceivable that a frugivorous primate might have developed a culture which would have been different from our own because of this difference in diet. Or, we might think of a primate that did not have stereoscopic, chromatic vision developing a culture. Or a species in which male dominance did not obtain, or in which the young were brought forth in litters, and so on. We might even imagine a culture produced by herbivorous quadrupeds, like Swift's Houyhnhnms in *Gulliver's Travels.* In short, our own culture, the one that has actually been realized, is not the only kind of culture that could have been produced. Discussion of the biological evolution of capacity for culture will depend, therefore, upon whether we wish to consider various conceivable kinds of culture or merely our own culture.

Clarence Day, in his charming little book *This Simian World,* has speculated upon the kind of civilization that a race of supercats, superants, or superelephants would have developed, and he compares these in turn with the civilization that a race of supersimians (man) has produced. Speculations of this sort are not merely idle or humorous; they require us to distinguish between the fundamental (symboling) and the incidental (stereoscopic vision, carnivorous-omnivorous diet). Spuhler makes it clear that he is addressing himself to the problem of the development of the culture "realized by members of the genus *Homo* and not to . . . all conceivable varieties of culture." Sahlins, also, observes that "if culture had not developed among a species of primates, but among animals of different behavioral characteristics, then the forms and development of culture would be basically different." None of the other authors concerns himself with this question.

It was the development of the ability to symbol that made possible the culture of the human species, and it would be difficult to imagine any other real species producing a culture that was not based upon this ability—although in popular science fiction one could easily imagine a species using electronic waves instead of symboling as the basis of a culture, but even here would not these waves have to have the properties of symboling?

Very little is known, or understood, about the biological evolution of the ability to symbol, as the preceding papers make clear (see also White, 1949). Gerard presents an excellent sketch of the process of neurological development that has culminated in this ability, but he is able to be no more specific than to note the large size of man's brain and the special development of certain of its local areas. The evolution of a large brain, large both absolutely and relatively, in man is unquestionably intimately related to man's unique ability to symbol, but the size of man's brain may be, and seemingly has been, exaggerated, as Spuhler points out.

Since articulate speech is the most characteristic form of expression of the ability to symbol, Hockett's analytical survey of systems of communication among mammals, birds, fish, and insects is illuminating in its suggestions as to how human speech may have been developed by the addition and combination of certain factors, most of which may be found in the communication systems of nonhuman species. According to Hockett, only one of his seven factors, that of duality of patterning, is found only in the articulate speech of man.

The significance of the evolution of erect posture and bipedal locomotion, accommodative vision, and the hand for the achievement of culture has been rather generally appreciated, and these points are competently covered in this symposium by Spuhler and Washburn. But the role of diet and the control, both cortical and social, of sexual activities in the origin and early development of human society and culture have been much neglected. The treatment of these points by Spuhler and Sahlins is noteworthy and illuminating.

Harlow's paper is interesting and informative in itself, but it has little bearing that I can see upon the problem of the origin of culture in the human species. He cites some factors—such as the ability to acquire and transmit patterns of behavior and the operation of cohesive forces that stem from affectional factors—as requisite to culture. But these are in no wise limited to primates but are possessed by other mammals and possibly by birds. But, in Harlow's thinking, these factors may provide the basis for the nonhuman cultures that he speaks of.

If, on the one hand, the evolution of anatomical structures has paved the way for, or has produced, certain kinds of behavior found in human cultural systems, the behavior of some of man's ancestors has had a profound effect upon the development of anatomical structures. Spuhler reasons that the use of hands, freed from the function of locomotion for extensive and varied use of tools, has had much to do with the development of the human brain and the ability to symbol. Washburn presents a persuasive argument for the effect of behavior, particularly the use of tools, upon the evolution of man's bodily structure: "It is probably more correct," he says, "to think of much of our structure as the result of culture than it is to think of men anatomically like ourselves slowly discovering culture." The use of tools has exerted a profound influence upon the evolution of man's bodily structure at many points, according to his thesis. Indeed, so pervasive and compelling has its influence been that one might well take Oakley's phrase literally: "Tools makyth man." Selection pressures, following upon the heels of erect posture and the use of tools, "changed the ape hand into the human hand." Selective pressures "coming in with the use of tools" brought about a "tripling of the size of the brain . . . after man . . . [became] a tool user. . . . from the long term evolutionary point of view, it is culture [does he not mean tool-using?] which creates the human brain." Since the chimpanzee "cannot learn to talk" because "the large amounts of brain necessary for speech are not there," it may be inferred that Washburn would think of the capacity for speech in the human species as a characteristic of a brain enlarged by tool using. The decrease in size of canine and incisor teeth in the evolution of man may also have been due to the use of tools, "perhaps of wooden tools which are long since gone."

Washburn goes so far as to suggest that "in the future it may be possible to demonstrate that the expansion of much of the cortex [in the course of the evolution of the human brain?] is directly related to new selection pressures associated with the evolution of complex social systems." Would he mean by this that the development of clan organization, or the eight marriage class system of the Aruntas, would *tend* to expand the cerebral cortex through "selection pressures"? Or that the vastly increased complexity of social organization in the United States since 1783 might have had this effect? If so, the possibility of demonstration does indeed lie "in the future."

Biological evolution among primates eventually produced human beings and culture. But once culture was achieved it opposed primate nature in a number of fundamental respects. It is Sahlins' essay that deals extensively and in a most illuminating manner with this point. Sahlins notes that there have been a number of continuities of behavior from prehuman primates to man, such as territoriality and dominance; but he very properly takes exception to such assumptions of continuity as the absurd equation of prostitution in human society with "the presenting behavior of a subordinate rhesus monkey."

Perhaps the most profound change in the behavior of primates in the transition from prehuman to human society was effected in the roles of sex and economics. In prehuman social organization, the sexual factor was preeminent; the economic factor subordinate, almost incidental. In the transition to human society the sexual factor became subordinated to the economic, mutual aid factor. This radical change was made possible and effected by articulate speech (the ability to symbol). Articulate speech made possible extensive and versatile cooperation among human beings whereas it was virtually absent among prehuman primates. Cooperation and mutual aid had a biological survival value and consequently they were developed and expanded. Sexual unions became a means of social organization for economic and other mutual aid ends. The contrast between human society and prehuman society in this respect is sharply brought out by the custom of wife-lending found among some primitive tribes: sexual rights and satisfactions are "waived in favor of socio-economic advantages."

One of the specific situations in which cooperation and mutual aid are expressed in human society brings out another sharp contrast between human and prehuman primates, namely, that with reference to food sharing. There is very little sharing of food among subhuman primates. A gibbon may indicate the presence of food to his group mates by vocal utterance, but food sharing—if we may use this term for the gibbon practise—seldom, if ever, goes beyond this. More often, in prehuman primate society, the dominant animal will take food, by force or intimidation, from the subordinate animal. But in primitive society, the sharing of food often goes beyond the rational requirements of economics and nutrition and becomes a social ritual whose function is to promote social solidarity—which, of course, has biological survival value.

This symposium goes considerably beyond the scope indicated by its title. It not only outlines the course of biological development that eventually produced man's bodily structure; it shows how primate behavior of various kinds, but particularly the use of tools, has profoundly affected the evolution of man's body and brain. The carry-over of certain trends and tendencies in prehuman primate behavior, both social and individual, and their expression in human behavior and culture, has been much discussed in the past, and this topic is competently treated in this symposium. But a point much more rarely considered is that which sees the emergence and establishment of culture opposing some fundamental traits of primate nature and practise. This, also, is illuminatingly treated. And, in addition to ideas, views, and interpretations, the various papers have gathered together and presented a wealth of factual information from far and near and offer it here for the reader in succinct, compact form.

LITERATURE CITED

WHITE, LESLIE A. 1949 The Science of Culture. Farrar, Straus & Co., New York (1958, paperback issued by Grove Press, New York).

WHITE, LESLIE A. 1959 The concept of culture. Am. Anthrop. In press.

Jacket designed by Richard Kinney
Set in Roman, Fairfield, and Neuland type faces
Printed on Allied Paper Company's Paperback Offset Paper
Bound in Riegel Paper Corporation's Carolina Cover.
Manufactured in the United States of America